On
Happiness

On Happiness

Epicurus

This edition published in 2020 by Arcturus Publishing Limited
26/27 Bickels Yard, 151–153 Bermondsey Street,
London SE1 3HA

Copyright © Arcturus Holdings Limited

AD006596UK

Printed in the UK

CONTENTS

INTRODUCTION

The life of Epicurus

Epicurus was born *c*.341 BC in the Athenian colony of Samos, an island in the Aegean Sea off the coast of present day Turkey. We know very little about his childhood. But according to ancient reports,* he took up the study of philosophy at fourteen 'because he was contemptuous of his school-teachers'. They were, it seems, unable to explain to his satisfaction what Hesiod says about 'chaos'.

The story, though perhaps apocryphal, is nicely illustrative. In Hesiod's account, before the origin of gods and nature, 'chaos' is all there is. Unlike the beginning of the Hebrew Bible, there's no intelligent creator-god in addition to the formless deep; when other divinities and the natural world arise, they are either generated out of chaos, or they arise spontaneously, as if from nowhere. It is rather hard to understand quite how it is supposed to work, and when young Epicurus presses on the issue, he is evincing an interest in comprehending the cause of the universe and the fundamental nature of the gods. But (like any good philosopher) he will not accept a view on the topic just because relevant authorities tell him it is true, nor because it is encapsulated in a beautiful poem, however deeply the poem had worked its way into the

---

* These quotations are from Diogenes Laertius (*c*.3rd century AD), *Lives and Opinions of Eminent Philosophers*. The translation used here (with some modifications) are from his biography of Epicurus in Inwood, Brad and L.P. Gerson (editors and translators), *Hellenistic Philosophy: Introductory Readings*, Indianapolis: Hackett Publishing Company, 1997.

fabric of Greek culture and intellectual life. The idea that in the beginning there was chaos, and that out of this chaos the world emerged, may be wonderful and awe-inspiring; but if the account doesn't make the genesis of the universe intelligible, and if we are given no good reason for believing that account in the first place, then there is philosophical work to be done.

A few years after dissatisfaction with his teachers drove him to philosophy, Epicurus spent some time in Athens, presumably to continue his pursuit of the subject. But, while he was there, the Athenian colonists were expelled from Samos, and Epicurus left Athens to join his father in Colophon, on the coast of modern Turkey. According to some accounts, Epicurus studied there with the philosopher Nausiphanes. It was a formative time for him intellectually. Nausiphanes was a proponent of the views of the pre-Socratic philosopher Democritus, who argued that all of nature is composed of atoms moving through the void, interacting with one another to produce the world that we see and experience. Democritus' atomism would come to have a profound and enduring impact on Epicurus' own thought.

In his early 30s, Epicurus left Colophon, and travelled to Mytilene (on the island of Lesbos) and then to Lampsacus (nearby on the mainland). It seems that, by this point, he had developed his own distinctive philosophical system; he began to teach, and gained a following among the cities' inhabitants. After five years (in 307 or 306 BC), he left to return to Athens. Perhaps he was driven out – his view that the gods are utterly indifferent to human affairs, and his strictly materialist conception of the soul, would certainly make that possibility credible. But perhaps he simply wanted to return to his parents' home, and to the centre of the Greek philosophical world.

In any case, he did return, this time settling in Athens for good. He purchased land just outside the city walls, and we are told that friends came from all around to live with him in 'the Garden' – as both the property itself, and the school located there, came to be known. It is perhaps no accident that Epicurus chose to locate his school outside the city itself; it's a fair representation of his attitude toward politics.

Given the meaning that now attaches to the word 'Epicurean', you might expect that life in the Garden involved a constant array of fine wine and gourmet food. If you've heard, in addition, that Epicurus and his followers were committed hedonists, you might even expect they lived profligate lives – endless feasts, drunkenness, free love, and the like. Anyone who showed up to the Garden with such expectations, however, was liable to be sorely disappointed. Epicurus himself claimed to be 'content with water and simple bread'. (The exception proving the rule is a request he sent to a friend, asking the friend to provide him with 'a little pot of cheese, so that I can indulge in extravagance'.) He expected his followers to be similarly abstemious. Diogenes Laertius reports that 'they lived very simply and frugally' in the Garden, and that 'generally their drink was water', though they occasionally allowed themselves a 'half-pint serving of weak wine'. A life subsisting on bread and water, where the occasional piece of cheese or glass of watered down wine constitutes the height of luxury: Epicurean hedonism might look more like asceticism. As we shall see, however, Epicurus argued not only that this involved no paradox, but that his was the only coherent, rational form of hedonism possible.

Such an austere way of life might give one the sense that Epicurus was a hard man, perhaps even something of a disciplinarian among his acolytes. But he reserved uncharacteristically exuberant praise for friendship – 'Friendship dances around the world, announcing to each of us that we must awaken to happiness' (p. 119) – and by all accounts this was reflected in the way he lived. We are told of his 'unsurpassed kindness to all human beings', and that 'his friends were so numerous that they could not be counted by entire cities'. Indeed, the Garden itself was more a community of friends, who shared a love for philosophy and some basic philosophical commitments, than it was a 'school' in any modern sense of the word. In this respect, their shared life in the Garden was, for Epicurus and his followers, a kind of alternative to the city-state whose politics they disdained.

There is one other feature of life in the Garden worth mentioning: It seems that the school admitted women (and slaves) as students. Although not unprecedented – we know of at least two women who studied at Plato's Academy – it is hard to express just how radical this would have been. Ancient Greece in general, and Athens in particular, was an extremely patriarchal society; aside from certain religious festivals, Athenian women were confined to domestic affairs. (The idea of a woman participating in politics would not have been scandalous so much as laughable.) In this context, it is hard not to see Epicurus' admission of female students as a political statement, whatever his aspiration to avoid politics.

Epicurus remained in Athens, gathering more disciples and developing his philosophical system, for over thirty years. In 271 or 270 BC, he died of kidney stones – a horrifically painful way to die, and yet in his last letter (p. 109) he tells his friend that the bodily 'pains and tortures' he undergoes are so outweighed by the 'joy of his heart at the memory of our conversations' that he counts his last day as 'blessed'.

Epicurean philosophy as a way of life

Epicurus did not see philosophy primarily as a theoretical study, though it certainly involved theoretical study. Philosophy was a way of life, whose goal was the achievement of 'tranquillity', freedom from mental disturbance and distress. We should, according to Epicurus, carefully inquire into the fundamental causes of natural phenomena only because, in doing so, we will be liberated from two very important sources of human worry. One is the belief that the gods are intimately involved in the world, tinkering with its natural workings and intervening in human lives for their own hidden purposes; the other is the belief that death is the greatest evil that befalls us. 'Mental tranquillity means being released from all these troubles' (p. 85).

How does Epicurus seek to release us? With respect to the gods, Epicurus (like Xenophanes and Socrates before him) draws attention to a tension or contradiction in the ancient Greek conception of the gods. Piety demands we take them to be perfect beings, paragons of virtue and self-sufficiency; and yet they are represented (in Homer, for instance) as constantly intervening in human affairs in ways that are frequently petty and cruel. But a perfect being would not act thus. Indeed, Epicurus reasons, given that the gods are self-sufficient and living blessedly happy lives, they have no motivation to pay any attention to our affairs at all. Some of us might find the prospect of divine indifference nearly as troubling as the prospect of divine punishment; but Epicurus suggests that it should free us to live our lives the best we can, without concern for propitiating the gods.

Similarly, Epicurus argues that, since the soul is material (as we shall see in the next section, everything is material) and will therefore be destroyed at our death, 'death is nothing to us' (p. 88). Once death has arrived, we will no longer exist to experience anything at all. Gods or no gods, we need fear no retribution in the afterlife; there is none.

Epicurus steadfastly maintains that this implies that it is irrational to fear death, and, given the equanimity he evinced on his deathbed, he seems to have meant it. Still, we might wonder about this argument. Even if I grant that death is the end, so that I cannot experience being dead, is it really irrational to fear the end of my ability to experience anything at all? Epicurus argues convincingly against an incoherent form such a worry might take – worrying about what it's going to be like to experience nothingness, or fearing that you will miss having experiences. But the fact that we won't be around to wish we were still alive might not dispel the hope to live long enough to see our projects through to completion. If so, it will not dispel the fear of death.

Epicurus, to his credit as a philosopher, does not suggest that we should try to convince ourselves that death is no evil anyway, since it would be a comforting illusion. Even if the goal is tranquillity, he refuses to get there by such means. He seeks always to provide compelling arguments, on the basis of fundamental principles for his basic commitments. But, given how little of his work has survived the millennia since his death (you've got a copy of pretty much all of it in your hands right now, aside from some fragments found in other authors), we often must read Epicurus as Epicurean philosophers ourselves. We must try to imagine how Epicurus would have completed an argument, or answered a worry raised about one of his central tenets. The arguments concerning the fear of death are a case in point. What we have from him on the topic raises difficult philosophical questions, and it is up to the reader to consider how they might be answered.

Epicurus also recognizes that arguments may not be enough, even if they are perfect as arguments. We might acknowledge that 'death is nothing to us', but go on fearing it; the fear may be instinctive, not immediately susceptible to rational argument. Epicurus thus counsels his followers to engage in a kind of therapy of the soul, by committing to memory the basic tenets of their philosophy, and rehearsing them almost as mantras, so that the tenets may become engrained in their souls. Thus, the first four propositions of his *Principal Doctrines* (p. 57) came to be known as the tetrapharmakos, the 'four-fold remedy', as if they were a kind of medicine.

Epicurean physics

The argument against the fear of death is only as good as the materialist theory on which it is based. Much of that theory derives from the pre-Socratic atomist Democritus, who had such a profound influence on Epicurus. Democritus held that everything that exists is composed of atoms shooting through 'the void' (i.e., empty space). The atomists accepted Parmenides's basic insight that nothing truly comes into, or goes out of, existence – the 'conservation of being,' you might say – but argued that we can account for the generation and destruction of natural entities, by supposing (1) that what most fundamentally exists are atoms, and (2) that generation and destruction (of, e.g., organisms) come about by the coalescing and dissipating of groups of atoms. The atoms themselves cannot be destroyed, because they themselves cannot be separated into their component parts – for they have no component parts. (Our word 'atom' is derived directly from the Greek word that Democritus used to characterize these basic building blocks of nature: atomos, meaning 'uncuttable'.) Everything, from the behaviour of celestial bodies to the growth of an olive tree to the mental life of human beings, was to be explained in terms of the interactions of these atoms as they move through empty space.

Epicurus took much of this physical theory on board, but he did not do so uncritically. He defended it against counter-arguments – many of which were levelled by the rather formidable philosophical opponent of Aristotle – often modifying the atomist theory in the process. Perhaps the most important objection to atomism was that it could not account for the purposive structure of natural entities. As Aristotle pointed out, biological organisms develop in such a way that their parts fit together to comprise a unified whole, each organ performing a function which enables the organism to survive as the kind of creature it is. Is it just a coincidence, Aristotle asks, that deer have teeth suited to chewing precisely the kind of food they need to eat? This objection would have been of particular interest to Epicurus, for the idea that the universe has a 'teleological' structure might seem to lend itself to the idea that things are guided by some divine hand (though Aristotle saw that it need not).

We find only the seeds of Epicurus' answer in his extant works, but we can glean his response from later Epicureans. The Roman Epicurean philosopher, Lucretius, writes that there was a time when the earth produced all sorts of creatures, most of them gruesome and ill-formed. The organisms we see today are suited to living on earth simply because they are the ones that survived in it. Of course, this may be more speculative than scientific, but it is worth noting that the Epicurean reply does have just the features needed to explain the appearance of purposiveness without postulating natural purposes: An element of randomness (for Lucretius, in the primordial past, when the earth generated all sorts of animals; for modern evolutionary theory, in genetic mutation), plus natural selection. Epicurus also maintains that there are infinite worlds. So, he has another reply available to him: Maybe it is a huge coincidence that organisms are well-adapted – but there's no surprise in that coincidence coming to pass if nature has infinite rolls of the dice.

Epicurus's epistemology

In the course of considering the variety of shapes and sizes of atoms, Epicurus suggests that as many 'differences of size' among the atoms should be admitted, as will enable us to explain 'the facts of feeling and sensation', but no more (p. 74). Setting aside the somewhat academic question about the range of atomic sizes, what is interesting is the kind of justification that Epicurus offers for his answer. Our scientific theory must be grounded, it seems, in 'the facts of feeling and sensation'. In saying this, he commits himself to what philosophers now call empiricism, the view that knowledge must be justified by appeal to what we experience or perceive.

This does not mean that Epicurus thinks we can only have knowledge of what we can see and touch. (We cannot see atoms, after all.) Rather, a good scientific theory is one which enables us to explain our perceptual experience of the world, though the explanation may involve postulating entities not themselves directly perceivable. Aristotle, who was himself an empiricist (however much he disagreed with Epicurus on virtually every other philosophical question), offers a helpful analogy: In foot races in ancient Greece, runners would sprint from the starting line, go around a pole at the end of the track, and sprint back, the race ending where it had begun. Philosophy, for the empiricist, works in a similar way. We begin with perception, proceed to develop a theory on its basis, but then we must 'return' to perception by seeking to explain it on the basis of our new theory. If we cannot make it back to perception, then our theory has failed.

How did Epicurus defend his adoption of an empiricist method? Sensation, Epicurus says, can never be in error, and that would seem to make it a very good basis for science. He does not mean, of course, that sensation never leads us to adopt false beliefs. His point is that falsehoods arise only because we make an over-hasty judgment on the basis of sensation; the sensation itself is never false. To take a common ancient example, let's say that I see a tower in the distance and judge that it is round, but the tower turns out to be square. According to Epicurus, my sensation was perfectly in order – that's just how square towers look from a distance – the mistake only came in when I made a judgment about it.

Readers familiar with the 17th century French philosopher René Descartes may be here reminded of his *Meditations*. Descartes suggested, that although one may doubt the existence of the very world, still I can be sure that there seems to me to be such a world. There is some similarity with Epicurus, no doubt, but the differences are equally important. Descartes thought of these indubitable appearances as being within the mind. Epicurus, however, is not speaking of an internal theatre, whose show we might be aware of even if there were no world at all. Epicurus means that it is true of the tower that it looks round to an observer standing at such and such a distance and such and such an angle.

Thus, for Epicurus, sensation is less a mental event than a natural fact, arising from the interaction between the world and a human being. In natural philosophy, our task is to develop a theory of the basic structure of the world that will adequately explain those natural facts.

Epicurean ethics

In ethics, Epicurus defended hedonism, the view that pleasure is the good at which all our actions should aim. Now, the idea that pleasure is good – that pleasure is desirable and worth pursuing, other things being equal – is hardly controversial (notwithstanding the protestations of certain Platonists and Puritans). Most of us accept that much, at least in practice; we often do things simply because they are pleasant to do, needing no further reason. Epicurus, however, is committed to something much stronger, that pleasure is the good: Pleasure is the one intrinsic value, in terms of which we should assess any other supposed good. On such a view, a human life is good to the degree it includes pleasure; a course of action is good insofar as it is likely to conduce to pleasure; a supposed virtue – courage, say, or wisdom – is good only if possessing or displaying the virtue is likely to produce more pleasure than pain.

The Garden was not the only hedonist game in town (though it proved far and away the more durable one); the Cyrenaics also maintained that pleasure is the good. Beyond this abstract point of agreement, however, Epicurus could hardly have differed more from the Cyrenaics. The Cyrenaics claimed that immediate sensual gratification was the greatest pleasure and therefore the proper goal of human action; they downplayed the pleasures of memory and anticipation, and even counselled against forgoing a present pleasure in the hope of thereby securing some greater pleasure in the future. Epicurus advises precisely the opposite, suggesting that we should engage in a kind of rational, hedonic calculation. Although pleasure is the good – or rather, because pleasure is the good – we shouldn't pursue every pleasure; we should 'pass over' any pleasure when a greater pain would result; and we should endure pain whenever it will conduce to pleasure in the long run (p. 90).

This would suggest divergent attitudes towards many activities – the Epicurean may recommend physical exercise and counsel against that last drink, for instance, while the Cyrenaic would have little grounds for doing so. But there is a further disagreement. Epicurus also criticized the Cyrenaic view that sensual gratification was a source of much pleasure in the first place. Sensual pleasures tend, according to Epicurus, to be accompanied or balanced out by pains. The greatest possible pleasure, paradoxical though it may sound, lies in a kind of neutral condition when we are free from both pain and pleasure. Thus, in another sense, the neutral condition is not really neutral at all, for (as long as one believes that the freedom from pain will continue) it involves a feeling of 'tranquillity' – and this, Epicurus contends, is the only condition which is perfectly, absolutely pleasant.

Epicurus does not reject sensual pleasures. Some, after all, stem from fulfilling a natural need; but even when they do not, it's perfectly natural to choose them whenever a greater pain would not result. But we need to be careful about indulging in such luxuries too much. Doing so might make us increasingly dependent on those luxuries, which could very well lead to anxiety down the road, when they are not forthcoming. It is for this reason that Epicurus counselled a kind of moderate asceticism among his followers, that such self-discipline might maximize their chances of enjoying the greatest possible pleasure – that of mental tranquillity. While the Cyrenaic is utterly dependent on the world to cater to his needs, the Epicurean thereby makes himself more self-sufficient, and thus godlike; by disciplining himself to take pleasure in simple things, he makes himself maximally immune to fortune's slings and arrows.

The legacy of Epicurus

Historically, Epicurus's natural philosophy and epistemology has been immensely important. Renewed interest in Epicurean atomism in the 17th century, for instance, contributed something to the development of the mechanistic conception of physics and thus to the scientific revolution. But in recent years, it is his ethical thought that has proved to be of particularly enduring interest.

In saying this, I do not mean primarily his view that pleasure is the only intrinsic good. That view was influentially taken up by Jeremy Bentham (1748–1832), the original proponent of Utilitarianism, though to quite different effect. Bentham argued that what we should care about is the aggregate pleasure of all those affected by any possible course of action, while Epicurus had focused on the pleasure of the individual agent. Bentham wanted to remain neutral on the question what pleasures are worth promoting; if someone would be pleased by the action in question, that fact goes into the hedonic calculating machine, no questions asked. As we have seen, by contrast, Epicurus recommended that we rigorously train ourselves to be able to enjoy certain pleasures rather than others.

These differences result from different conceptions of the nature and role of philosophy, and thus bring out precisely why philosophers have recently been looking back, past Bentham and Hobbes, to Epicurus and the other Hellenistic schools; so that the comparison with Bentham brings out, surprisingly, not so much continuity as contrast. For decades, and until fairly recently, much moral philosophy in the English-speaking world was concerned almost exclusively with the question of whether we should prefer the Utilitarian or Deontological theory of right action. Certainly, that is an important and interesting topic, but it did have a stultifying effect, leaving moral philosophers to occupy themselves with the task of inventing increasingly ingenious counter-examples to one moral theory or another. It was in this context – when moral philosophy seemed more and more to have become an academic exercise, only occasionally making contact with our lives – that the Epicurean conception of philosophy as a way of life re-emerged as an important corrective. Our world is certainly a very different place than the one in which Epicurus lived, and there is much that a contemporary reader will find strange and perhaps even repugnant in what he says. And yet the basic spirit of the enterprise, of engaging in philosophy as a kind of therapeutic work on the self, continues to inspire.

Nicholas Gooding,
University of California, Berkeley

THE PHILOSOPHY OF EPICURUS

ROBERT DREW HICKS

(*selections from Robert Drew Hicks*, Stoic and Epicurean)

EPICURUS AND HEDONISM

It is now time to inquire into the nature of Epicurus's teaching, which met with such an enthusiastic reception and was greeted almost like a revelation. Philosophy was defined by Epicurus as 'a daily business of speech and thought to secure a happy life.' Here is struck the note of intense earnestness characteristic alike of Epicurus and his age. Philosophy is a practical concern; it deals with the health of the soul. It is a life and not merely a doctrine. It holds out the promise of well-being and happiness. This is the one thing needful. Literature, art, and the other embellishments of life are not indispensable. The wise man lives poems instead of making them. Accordingly, as we have seen, Epicurus regarded with indifference the ordinary routine education of the day in grammar, rhetoric, dialectic, and music, and for mere erudition he had a hearty contempt. The only study absolutely necessary for a philosopher was the study of nature, or what we now call natural science, and this must be cultivated, not for its own sake, but merely as the indispensable means to a happy life. Unless and until we have learned the natural causes of phenomena, we are at the mercy of superstition, fears, and terrors.

We must defer to a subsequent chapter the consideration of the steps by which Epicurus was led to the conclusion that the external world is a vast machine built up by the concourse of atoms in motion without an architect or plan. Suppose, however, this conclusion firmly established; what has our philosopher to tell us respecting human life and action? In what consists the happiness which is our being's end and aim? This had, by the time of Epicurus, become the chief question of philosophy, and, strange as it may appear, the answer is no new doctrine, but one which had often been proposed and discussed in the ancient schools. He identifies happiness, at least nominally, with pleasure, and he means the pleasure of the agent. His is a system of Egoistic Hedonism. Verbally, then, he is in agreement with Aristippus, the founder of the Cyrenaics, with the Socrates of Plato's *Protagoras*, and with Eudoxus, whose doctrine of pleasure is criticized by Aristotle in the *Nicomachean Ethics*. The same doctrine is discussed in more than one of Plato's dialogues, sometimes apparently with approval, sometimes with disapproval. The historical Socrates never, so far as we know, reached a final definition of Good. He knew no good, he said, which was not good for somebody or something. His teaching would serve equally well as an introduction to Egoistic Hedonism, to Universal Hedonism, to Utilitarianism, or to Eudaimonism. The difficulty at once occurs; if pleasure and good are identical, why is it that some pleasures are approved as good and others condemned as evil? Why, on this hypothesis, should life ever present conflicting alternatives in which we are called upon to choose between doing what is good and doing what is pleasant?

Every hedonistic system must face this problem. Some progress had been made by Plato in the Protagoras.

There his spokesman, Socrates, maintains that since every one desires what is best for himself, and since he further identifies good with pleasure, evil with pain, he avoids pleasure when it is the source of still greater pain, and only chooses pain when a greater amount of pleasure results from it. In this Epicurus heartily concurred. He never recedes from the position that pleasure is always a good and pain always an evil, but it does not follow that pleasure is always to be chosen, pain to be shunned. For experience shows that certain pleasures are attended by painful consequences, certain pains by salutary results, and it is necessary to measure or weigh these after-effects one against the other before acting. 'No one beholding evil chooses it, but, being enticed by it as by a bait, and believing it to contain more good than evil, he is ensnared.' We now get a clearer notion of the end of action, which turns out to be the maximum of pleasure to the agent after subtraction of whatever pain is involved in securing the pleasure or directly attends upon it. At this point Epicurus parts company with Aristippus, whose crude presentation of hedonistic doctrine identified the end with the pleasure of the moment.

So soon as conditions and consequences are taken into account, pleasure tends to become an ideal element capable of being realized in a series of actions, or in the whole of life, but not to be exhausted at any given point of the series. More important, however, for determining the exact significance of this conception is the incursion which Epicurus makes into the psychology of desire. Desire is prompted by want; unsatisfied want is painful.

When we act in order to gratify our desires, we are seeking to remove the pain of want, but the cessation of the want brings a cessation of mental trouble or unrest, and this must carefully be distinguished from positive pleasure which is itself a mental disturbance. Experience shows a succession of mental disturbances, painful wants, the effort to remove them, and the pleasurable excitement which attends their removal. But all this shifting train has for its natural end and aim a state which is neither want nor the pleasurable excitement of satisfying want. All of them are fugitive states as contrasted with the resultant peace and serenity in which they end. The former agitated sea, whether swept by storms or tempests in gentle, equable motion, the latter to the profound calm, waveless and noiseless, of a sheltered haven. Beyond this neutral state of freedom from bodily pain and mental disturbance it is impossible to advance. We may seek new desires; we are only returning to the old round of painful want, desire, and pleasurable excitement of removing the want. There is only one way to escape from this round, and that is to be content to rest in the neutral state. After all, this is the maximum of pleasure of which we are capable: any deviation from it may vary our pleasure but cannot increase it. 'The amount of pleasure is defined by the removal of all pain. Wherever there is pleasure, so long as it is present, there is neither bodily pain nor mental suffering, nor both.' The consideration of these elementary facts should regulate preference and aversion. Prudence demands the suppression of all unnecessary desires. Epicurus does not carry renunciation so far as the Buddhists, who hold that to live is to suffer, and explain the will to live as that instinctive love of life which, partly conscious, partly unconscious, is inherent in all living beings. They look for their rest in Nirvana. Certain things, says Epicurus we must desire, because without them we cannot live, and life to Epicurus is worth living; and yet the repose which consists in the cessation of desire is, after all, not altogether unlike the Nirvana of the Buddhists.

In this negative conception of happiness as freedom from pain, whether of body or mind, Epicurus must have been influenced by the ethical teaching of Democritus, who also made happiness in its essential nature consist in the cheerfulness and wellbeing, the right disposition, harmony, and unalterable peace of mind which enable a man to live a calm and steadfast life. Democritus also exalted mental above bodily pleasures and pains, and laid stress upon ignorance, fear, folly, and superstition as causes of those mental pains which tend most to disturb life. With Epicurus the great obstacle to happiness is neither pain nor poverty, nor the absence of the ordinary good things of life; it is rather whatever contributes to disturb our serenity and mental satisfaction, whatever causes fear, anxiety – in a word, mental trouble. To be independent of circumstances is his ideal; that a man should find his true good in himself. He is ready with practical suggestions for realizing this independence. Groundless fear must be removed by the study of nature, which shows that the fear of death, the fear of the gods, belief in Providence and in divine retribution are chimeras; desire must be regulated by prudence and the virtues cultivated as the indispensable means to a pleasant life .Fatalism is not true any more than the doctrine that all things happen by chance. The future is not in our power;

our actions alone are in our power to make them what we please. The letter to Menoeceus sets forth the ethical doctrine of Epicurus in a convenient summary (see page 91).

In this document, scientific ethics, as the term is now understood, is overlaid with a variety of other topics. The practical exordium, the dogmatic inculcation of moral precepts, the almost apostolic fervour and seriousness of tone find their nearest counterpart in the writings of religious teachers. We are reminded by turns of the Proverbs of Solomon and of the Epistles of St Paul. The rejection of the popular religion and the denial of divine retribution are coupled with an emphatic affirmation of the existence of blessed and immortal gods. The instinctive fear of death is declared to be groundless; and here the writer enlarges upon a theme, first started by the sophist Prodicus, that death is nothing to us. Incidentally, the value of life is vindicated and the folly of pessimism exposed. The limitation of desire is seen to involve habituation to an almost ascetic bodily discipline, in order that the wise man may become self-sufficing, that is, independent of external things. Lastly, the freedom of human action is stoutly maintained in opposition to the doctrine of natural necessity first promulgated by the earlier Atomists Leucippus and Democritus, but at the time of Epicurus developed with the utmost rigour and consistency by the Stoics. On the main question there is no uncertainty. The pleasure of the agent is the foundation upon which Epicurus, like many after him, sought to construct a theory of morality which would explain scientifically the judgments of praise and blame passed by the ordinary man. All systems allow that there are self-regarding virtues and self-regarding duties, and when he has given his peculiar interpretation of pleasure, Epicurus has no great difficulty with these. But the case is different when we come to the social virtues and the duties which a man owes to his neighbour. In a system which makes self-love the centre of all virtues, and in which all duties must be self-regarding, if we accept, as he did as a psychological truth that by instinct and nature all are led to pursue their own pleasure and avoid their own pain, how can any conduct savouring of disinterestedness find rational justification? This was the great problem of the English and French moralists in that age of enlightenment, the eighteenth century. As then, so two thousand years before in Greece, extreme individualism was the order of the day. The primary fact is individual man as he is given by nature, and all that lies outside this, all that he has been made by institutions like the family and the state, all the relations that go beyond the individual are subsequent, secondary, derivative, requiring to be explained from him and to justify their validity to the reason. Take a concrete instance. Whence came the rules of justice? What makes actions just and how is my obedience to such rules, enjoined by Epicurus, an indispensable means to my own happiness? In short, how does disinterested conduct arise under a selfish system? The answer given to this question was often repeated later. It reappears in Hobbes and Rousseau. Before dealing with it, it is necessary to consider briefly the Epicurean conception of the growth of human civilization from the earliest times.

Epicurus on the origins of political communities and the demands of justice

Looking back at the past history of our planet, Epicurus derives all organisms, first plants, then animals, from mother earth. The species with which we are familiar are those which, being adapted, to their environment, prospered in the struggle for existence. They were preceded by many uncouth creatures and ill-contrived monsters, many races of living things, which have since died out from lack of food or some similar cause. Apart from the undeniable suggestion of one feature in the doctrine of evolution, the account of the origin of life is in its details wholly unscientific and even repulsive. But with surer insight primitive man is described as hardier than now: destitute of clothing and habitation, he lived a roving life like the beasts with whom he waged ceaseless warfare, haunting the woods and caves, insensible to hardship and privation. The first step in advance was the discovery of fire, due to accident. Afterward man learned to build huts and clothe himself with skins. Then the progress of culture is traced with the beginnings of domestic life through the discovery and transmission of useful arts. As comforts multiplied, the robust strength of the state of nature was gradually impaired by new disabilities, particularly susceptibility to disease. Language was not the outcome of convention, but took its rise from the cries which, like the noises of animals, are the instinctive expression of the feelings and emotions. Experience is the mother of invention and of all the arts. They are all due to the intelligent improvement of what was offered or suggested to man by natural occasions. None of the blessings of civilization are due to the adventitious aid of divine agency. Man raised himself from a state of primitive rudeness and barbarism and gradually widened the gulf which separated him from other animals. From the stage when men and women lived on the wild fruits of the wood and drank the running stream, when their greatest fear was of the claws and fangs of savage beasts, to the stage when they formed civic communities and obeyed laws and submitted to the ameliorating influences of wedlock and friendship, all has been the work of man, utilizing his natural endowments and natural circumstances. Religion has been rather a hindrance than a help in the course of civilization. Next to the use of money, the baleful dread of supernatural powers has been the most fruitful source of evil.

In this historical survey, where shall we find the origin of law and justice? Epicurus was fully convinced that in the present state of society 'the just man enjoys the greatest peace of mind, the unjust is full of the utmost disquietude'; and yet injustice is not in itself an evil, and in the state of nature man is predatory.

Why should the wise man observe it if he find secret injustice possible and convenient? Epicurus frankly admits that the only conceivable motive which can deter him is self-interest, the desire to avoid the painful anxieties that the perpetual dread of discovery would entail. Even if the compact could be evaded, prudential considerations forbid it, since the risk of detection is enormous and the mere possibility of discovery is an ever-present evil sufficient to poison all the goods of life. That such motives do not weigh with criminals is irrelevant; we are dealing now with the wise and prudent man. Justice, then, is artificial, not natural.

In other words, justice is the foundation of all positive law, but the positive law of one state will differ from that of another. Thus a law judged to be inexpedient is no longer binding. The old sophistical quibble that no positive law can be unjust Epicurus, from his stand-point, can easily expose, and he is equally well able to meet the conservative dislike and dread of legislative innovation as something essentially immoral.

Thus civilization is an advance upon the condition of primitive man; nor does Epicurus ever contemplate the possibility of undoing what has been done. Applying the standard of human good in his own conception of it as tranquil enjoyment, he pronounces government to be a benefit to the wise so far as it protects them from harm. But it does not therefore follow that they should themselves take part in political administration; they are only advised to do so in circumstances where it is necessary and so far as it is necessary for their own safety. Experience shows that as a rule the private citizen lives more calmly and safely than the public man. The burdens of office are a hinderance rather than an aid to the end of life. 'The Epicureans,' says Plutarch, 'shun politics as the ruin and confusion of true happiness.' An unobtrusive life is the ideal. To strive at power without attaining one's own personal security is an act of folly certain to entail lasting discomfort. Restless spirits, however, who cannot find satisfaction in retirement are permitted to face the risks of public activity. To all forms of government the Epicureans were theoretically indifferent, but the impossibility of pleasing the multitude and the necessity of strong control inclined them to favour the monarchical principle. Under all circumstances they recommended unconditional obedience. The traditions of the old republican life of petty Greek states demanded from the citizen far more than this active co-operation, personal sacrifice, enthusiasm for the common cause. Judged by this standard, the Epicurean would seem to take an unfair advantage of the state. He got all the protection it afforded and shirked as much as he could of its burdens. But, in reality, what he was prepared to contribute would fully satisfy the demands of the modern territorial state. To obey the laws, to pay taxes, to assist by an occasional vote in the formation of public opinion constitutes nowadays the whole of civic duty for the vast majority of citizens. Under existing conditions how can it be otherwise? For, in order to integrate, as it were, these multitudinous infinitesimals, organization is required; but division of responsibility and specialization of function circumscribe personal effort. Again, when the popular cry has been adequately voiced by press or platform and has taken effect through proportional representation or other constitutional means, the greatness of the results secured and the very perfection of the machinery for securing them leave less and less scope to private initiative.

The consistent application of individualist principles might enjoin a severance, so far as is possible, from the ties of the family no less than of the state, and the picture of the wise man represents him as shirking these responsibilities also. But such a counsel of perfection has regard to special circumstances, and in all fairness the actual conduct of the man should be allowed to correct the supposed tendency of his system. Now, by his kindness to his brothers, his gratitude to his parents, and his tender solicitude for his wards, Epicurus is proved to have cherished warm family affection himself. Nor is it reasonable to presume that the philosopher who deprecated suicide, except in extreme cases, and set the example by so cheerfully enduring severe physical pain, can ever seriously have intended race suicide.

Political association, even if originally based upon a contract, has its present sanction in pains and penal ties. It is at best a compromise, a *pis aller* of only relative and subsidiary value. Men submit to the compulsion and constraint which it entails for fear of finding something worse. The true form of association is that in which man surrenders nothing of his original freedom, and this Epicurus believed to be realized in friendship, upon which he set the highest value. The only duties that Epicurus recognizes are those voluntarily accepted on reasonable grounds, not from natural instinct or compulsion of circumstances.

The terms in which it is extolled recall the eulogies lavished upon the Christian grace of charity or love. It was the signal characteristic of the little society in the founder's lifetime, and it continued a prominent trait of the sect to the latest times. Upon its own principles no ethical system which starts with self-love can recognize disinterested conduct. Nor did Epicurus anticipate Hume's discovery and call in sympathy as a necessary supplement to self-interest. He is, therefore, obliged to maintain that friendship, like justice, is based solely upon mutual utility. The services rendered have the same selfish motive which prompts the farmer to commit the seed to the soil in expectation of a future harvest. So alone the theory is consistent; friendship, like the cynic's gratitude, must needs be a lively sense of favours yet to come. There is, of course, a difficulty at the beginning. Someone must make the start. Benevolence would cease to be a virtue if it ceased to be self-regarding. Yet it was upon this unsound basis that devoted friendships were based. When we are told that the wise man will, upon occasion, even die for his friend, the suggestion of disinterested action, however inconsistent, can hardly be dismissed.

In the foregoing sketch the main questions of ethics have come before us and the answers of Epicurus have been indicated in outline. Like his rivals the Stoics, he made his appeal to the world primarily as a moral teacher, an inquirer whose aim was to deal comprehensively and systematically with moral problems. To this inquiry the study of nature, which will occupy us in the next chapter, was subordinate. He had convinced himself that the main fruit of philosophy consisted in happiness of life and that philosophy was successful just in so far as this was promoted. This aspect of the system will become more apparent if we now consider the remarkable collection of its more important tenets, which has come down to us in the form of some forty isolated quotations from his voluminous writings.

Whether Epicurus himself made this collection or whether it was formed by his disciples cannot now be precisely determined. At a very early time it obtained a wide circulation among his followers, who were ever afterward recommended to commit to memory this collection of golden maxims as well as other shorter or longer epitomes of the master's teaching. The importance attached to these authoritative pronouncements must be our excuse for reproducing the greater part of them, although it will be obvious that except the first, which lays the foundation for his views upon religion, and the twenty-second, twenty-third, and twenty-fourth, which deal with his theory of knowledge, they are of an ethical character and must therefore simply recapitulate the ethical theory which we have already attempted to expound.

It is worthwhile to make an effort to discover the real Epicurus, to understand what manner of man he was. Our best materials are his own writings. The letter to Menoeceus has already been translated; that to Herodotus will occupy us in the next chapter, perhaps to the weariness and impatience of the reader. These letters together with other fragmentary records certainly convey the impression of a strong personality. We see that Epicurus had a logical mind, was a great systematizer, belonged, in short, to the class of daring and self-confident innovators. Like others of this class, he felt that he had a mission, and under great difficulties, in face of much opposition, laboured with unremitting industry to accomplish a self-imposed task.

EPICURUS AND ATOMIC THEORY

Among conflicting theories, Epicurus chose to stand by the mechanical conception of the physical universe, when it had fallen into disfavour, and unhesitatingly rejected the fashionable teleology. His doing so testifies to his acute intellect and critical insight, but still more to the honesty, fearlessness, and independence with which he invariably followed his convictions. He also popularized the system he adopted and lent it a new lease of life. So much will be readily admitted, but an impartial estimate of his services cannot go beyond this. He made no discoveries in science himself, nor did any Epicurean after him. He rather discouraged the prosecution of physical inquiries of any sort beyond a certain point. His attitude to natural science as a whole deserves careful consideration. He takes it up because, if we are to be happy, we must be released from mental trouble, above all from groundless fears, more particularly the terrors of superstition, the fear of the gods, and the dread of death. Without this strong impelling motive Epicurus would never have engaged in the study of nature at all. His sole aim is to convince himself that these terrors are unreal and imaginary, and if, incidentally, he discovers a great deal about the constitution of the world and man's place in nature, it is because he cannot otherwise banish these terrors from the mind. Scientific investigation is permissible only so far as it conduces to this end by laying down the true place of man in the system of things. Beyond this there is no need to go. The laboratories, museums, observatories, and other appliances of modern times for research and discovery, would thus be condemned in anticipation as superfluous. Knowledge in itself and for its own sake he regarded as of little worth. And this was no mere passing phase; it expressed the man's fundamental and settled conviction.

The indivisible atom was the basis of all Epicurean physics. It seems highly probable, then, that Epicurus himself would incline to the assumption of an indivisible unit of length, a sort of materialized point. If this surmise be correct, he found himself at variance with what we may call the orthodox school of geometers. Their fundamental notions of line and point he could not accept, and, as they were involved in the whole of geometry, he would feel bound to condemn the science as false. As will hereafter be seen, there is some evidence that he did not altogether accept the continuity of motion, but rather resolved it into a series of progressions, each taking place in an instant of time over an indivisible unit of space. His denial of continuous corporeal magnitude would of itself suffice to bring him into collision with the mathematicians; and this hostility would be strengthened if he also inclined to regard space, time, and motion as in the ultimate analysis not continuous, but discontinuous as made up of discrete minima.

Epicurus' empiricist theory of knowledge

But, be this as it may, it is high time to inquire what scientific principles, if any, our philosopher admitted. He was certainly no sceptic. He did not hold that every statement is uncertain, because as much can be said against it as for it, and, as a necessary consequence, that all science is founded on nothing better than probability. On what general principles, then, did he conceive himself entitled to assert or believe anything? This inquiry, preliminary to his physics, he himself entitled Canonic, because it dealt with the canon or rule of evidence. First, every statement must relate to what is given, to facts or phenomena. Epicurus is not concerned with the grounds on which from one proposition we infer another, but with the far more fundamental question: On what ultimate grounds is a statement of fact based? All phenomena are either immediately certain or not, and it is possible to pass from the one region where there is immediate certainty to the other region, which is not thus immediately certain; in other words, from the known to the unknown. Such a process is analogous to the modern induction. For deductive logic, the theory of the syllogism and definition, Epicurus had the utmost contempt. On the other hand, the few general and preliminary remarks of which his Canonic consists contain the germs of a thoroughgoing inductive logic.

The Epicurean theory of the universe is built upon this foundation. The existence of the phenomenal universe is everywhere assumed. Things exist outside us. We know them only through sense, which alone gives a conviction of reality. This conviction of reality attaches not only to the external objects which are perceived, but with equal strength to the internal states or feelings, especially the feelings of pleasure and pain of which we are conscious. All true belief and assertion, then, must be founded upon our sensations and feelings. What we immediately perceive and feel, that is true.

It is only through sense that we come into contact with reality; hence all our sensations are witnesses to reality. The senses cannot be deceived. There can be no such thing, properly speaking, as sense-illusion or hallucination. The mistake lies in the misinterpretation of our sensations. What we suppose that we perceive is too often our own mental presupposition, our own over-hasty inference from what we actually do perceive. When we see an oar which is half immersed in water appear bent, the image or film which reaches the eye is really bent, but the judgment of the mind that the oar itself is bent is no part of the perception, it is a gratuitous addition to it. The mind confuses two quite distinct processes or movements, the perception which is infallible, and the conscious or unconscious inference from it, which is after all mere presupposition or opinion, a groundless belief. The region of certainty, then, confined as it is to the direct presentation of sense, is even so by no means as extensive as we might at first suppose. Sensations themselves must be scrutinized, and the element which the mind itself has added must be removed before we get back to the original data, the perceptions which put us in touch with reality.

Turning now to the other and vaster region of the unknown, which is not accessible to direct observation because sensation is strictly limited to here and now, we observe that some part of it may hereafter come within our ken and be directly observed. This Epicurus denotes as 'that which awaits confirmation.' Cognition is an interrogative process. We put the question and wait until experience and reality, under favourable circumstances, supply the answer. But our knowledge, confined within these limits, would be very inadequate. By what we have above called an inchoate induction Epicurus regulates the steps by which we anticipate all experience with certainty.

His fundamental assumption is the uniformity of experience: that whatever occurs in the sphere beyond knowledge must follow the same laws of operation as what is known to occur within the range of our experience. It is right, then, to affirm about the unknown (i) what is confirmed and witnessed to by the known, or at least (ii) what is not directly witnessed against by the known. Thus the criterion, the supreme test of validity, is future experience, experience repeated or, at all events, not contradicted. The second half of this canon is by no means so sound as the first. It is capable of wide application, and must allow many doubtful explanations to pass for matters of belief. What is the ground on which Epicurus believes that there is an infinity of worlds, that the blessed and immortal gods inhabit the *intermundia*, that films from external objects enter the sense-organs and the mind, thus causing sensation and thought propositions for which there is not a tittle of positive evidence? His reply is: 'Nothing that we know by direct observation contradicts any one of these assertions.' And so Epicurus gives them, we may say, the benefit of the doubt.

Another caution is needed. If reasoning is to be anything better than mere quibbling, special attention to language is necessary. Every term that is used must call up a clear and distinct conception or idea, which again must be based upon one clear and distinct perception. To general terms correspond not single images, but the resultant of an accumulated series of images, the individual peculiarities of which are blunted and fused in a single pictorial type, much in the same way as when the photographs of different individuals are superposed on each other in order to form a composite photograph. But every perception in the series must be clear and distinct, in order that the resultant may have these qualities. In this way we obtain what Epicurus called 'preconceptions,' which take their place beside perceptions and feelings. They are the nearest approach which his system allowed to general notions. When a general term like 'man' is used, it calls up to the mind the preconception of man, the generic type in which the images of particular men are fused and blended. With this explanation and qualification we may even be permitted to substitute 'general notion' for 'preconception,' always remembering that it is an inexact equivalent. It remains to explain what precisely Epicurus understood by reasoning in which general terms are used, and what part it plays for him in the acquisition of knowledge. Sense gives us the raw material of knowledge in trustworthy perceptions and internal feelings, but he never denied that we also attain knowledge by the exercise of reason. Indeed, all the more important propositions in the general theory to be hereafter unfolded are attained by its aid. Reason or reasoning is to him a mental operation, which deals, not with particular things, but with generic types or notions. If our knowledge did not go beyond sensation it would consist in isolated, particular facts. In that case it would be difficult, if not impossible, to make the inductive leap from the known to the unknown. Reasoning, then, is the application, in a region where direct observation fails, us, of preconceptions or general notions derived from sense, their validity being guaranteed by repeated and uncontradicted experience. But future experience is the sole criterion by which all our reasoned conclusions must be tested. The great doctrine

modern astronomer. To the latter the universe is resolved into countless suns, each with its attendant planetary system, and the nebulae out of which such solar systems are believed to have developed. For him the many suns and planetary systems are dotted here and there throughout space, as were the 'worlds' of Epicurus.

The next division of the subject is concerned with the manner in which we are affected by external objects, and we begin with a remarkable hypothesis, that from the exterior surfaces of all composite bodies there is a perpetual emission of particles of matter or what we may call 'films.' Epicurus describes the enormous velocity with which they must travel in order to reach us, as in his view they appear to do, instantaneously. This, however, in no way detracts from the importance of these almost parenthetical remarks about motion; not the motion of atoms, which is at all times uniform, but the motion of systems of atoms. What is here said applies to all such systems, whether the union is loose and easily broken, as is the case with an invisible film, more close as with the air and other gases, closer still as in water and other fluids, or comparatively permanent and durable as in earth and the various composite bodies which we call solid. In all cases alike the system moves slowly if resistance is encountered, either externally from the medium, air or water, or internally and this is far more important from the jostling, collision, and backward rebound of the single atoms composing the system. Such internal resistance tends to impede the system. So, also, would the pause of rest, if the system reached a point, stopped, and then went on. But this, he explains, the film does not do unless it encounters resistance.

In sensation an image strikes upon the sense-organ. In every act of preconception or of memory an image strikes the mind. A series of repeated images or the traces which they leave behind them in us produce a presentation of the shape or properties of the external object from which they came. And if the presentation be obtained in this way by direct contact, whether on the senses or the mind, it corresponds exactly in shape and properties with the external object. If these conditions are fulfilled, the shape as presented to us in sensation and memory or in pre-conception is the real shape of the object, the proper ties so presented are the very properties which the external object has. Epicurus is here passing from the subject of films in general to the veracity of the reports of the senses. A theory of mediate perception must answer the question: How do I know that what I receive through the medium is an exact copy of the object?

Atoms, then, streams of atoms emitted from the surface of composite bodies, are the causes of our perceptions of external things. The things perceived have colour, sound, and odour. Is this so with the atoms?

The atom is unchangeable *ex hypothesi*, and this may be secured provided that the qualities which the atom possesses are themselves unchangeable. So long as the shape remains unaltered through all the motions, collisions, and entanglements which befall the atom, since there is no void within it, there will be no alteration in size and, since weight depends upon size or mass, there will be no alteration in weight. In this way size and weight may be regarded as properties necessarily conjoined with shape. Neither of them would be affected by different arrangement or position of the atoms, on which ultimately depend the qualities which composite bodies have and atoms have not. Take colour. In a composite body or aggregate of atoms differently placed and arranged and, it may be, themselves different in shape and size, the colour which we perceive as belonging to this aggregate, and which by the canon of Epicurus really does belong to it, is a consequence of these same atomic positions, arrangements, shapes, and motions, and a change in them may change the colour of the thing or composite body without that thing necessarily ceasing to be what it was. The question may be asked: To which division of qualities does colour belong? Is it a property, a *coniunctum*? Or is it an accident, an *eventum*? It seems safest to reply that generic colour, colour of some sort or other, is a property of all visible things, so long as they are visible; but particular colour is an accident or *eventum* of a particular visible thing, which often changes like the hues of a sunset cloud or in a peacock's tail, owing to the difference of atomic motions produced by light or some other external influence; lastly, that when a body ceases to be visible it has no colour. The qualities which are not inherent are accidental qualities, *eventa*, such as whiteness, triangularity, which a thing may gain or lose without ceasing to be what it is. Figure or shape in general, however, is not such an *eventum*, but an essential property, or *coniunctum*, of all material things whether visible or not. We regard shape as something which a material thing must have as long as it exists at all. We recognize that the shape changes, but we still think of the thing as being the same under an altered shape, as in the growth of animals and plants or when the same block of wax is moulded into different shapes. In other words, so long as a material thing persists it must have some shape or other.

From the Letter to Herodotus: '[N]ot only must we accept as impossible subdivision *ad infinitum* into smaller and smaller parts [since the natural world is comprised of indivisible atoms]...but in dealing with finite things we must also reject as impossible the progression *ad infinitum* by less and less increments.' The notion of such a progression is the groundwork of the famous puzzle of Achilles and the tortoise, propounded by the Eleatic Zeno. Achilles, who runs ten times as fast as the tortoise, gives the latter a start of a metre. When Achilles has run one metre the tortoise is one decimetre in advance; when Achilles has got as far as this he finds the tortoise a millimetre in advance, and so on *ad infinitum*; whence Zeno wished it to be inferred that Achilles will never overtake the tortoise. Epicurus simply denies the possibility of continuing *ad infinitum* such a progression, formed by a series of increments, each term in the series being a definite fraction of the preceding term, precisely as he denies the possibility of continuing *ad infinitum* the process of subdivision of a finite body, e. g., by taking half, then the half of this half, or one-quarter, next the half of this quarter, or one-eighth, and so on. The latter series of fractional divisions is the complement of the former, that of fractional increments. The impossibility in the one case and in the other is bound up with Epicurus' assumption that in the last resort not only body, i.e. matter, but the dimensions of body, which are conceived as traversed in motion, are discrete. To the atom, the indivisible minimum of body, corresponds an indivisible minimum of a dimension, of spatial dimensions, length, breadth, and depth, at any rate when the space is filled and occupied with body, under which conditions alone we have the clear evidence of sense and intellect for progression from point to point upon it.

Epicurus takes the doctrine to imply that the number of atoms in each thing is infinite, and he objects that, however small in size the individual atoms, an infinite number of them could produce a body not finite but infinite. His second objection is that, if the atoms be of finite size and an infinite number of them be contained in a single thing, the progression from the extremity of the first to the extremity of the next, and so on to that of the last, would be a never-ending progress, which he has before declared to be impossible. The word translated here 'extremity' and in Lucretius 'cacumen' will best be understood if we take an angular point or projection or extreme edge on any sensible body of finite size, e. g., the 'point' of a sharpened lead pencil or the corner of a cube. If each atom has a certain shape it must be conceived on the analogy of finite bodies to project some part of this shape which the mental vision can distinguish. But what, it may be asked, of spherical atoms? As it is impossible to see the whole of a finite sphere with the bodily eye or to present to the eye of the mind the whole of a spherical atom at once, the part which we do see will be bounded. The outside or edge in the part we do see is in this case the extremity projecting into view. This applies to the visualized pictorial image as well as to actual perception.

Epicurus regarded a line or length as made up of certain minima of length, his substitute for the geometrical point. Geometers denied that a line could be conceived as made up of, or could be resolved into, a series of points. But in their conception and definition of a point they differed widely from Epicurus. The geometers assumed infinite divisibility; there was a point wherever the line could be divided. Epicurus introduces us to discrete minima of length which bound finite perceptible lengths precisely as the geometer's points bound his lines.

These minima cannot first exist apart and then, in virtue of possessing the attribute of motion, unite together to form the atom. Our pressing business now is with the atom conceived on the analogy of finite bodies as occupying space and therefore extended, and, being extended (or, as Epicurus prefers to say, 'capable of being traversed'), as having parts. We must not by one whit modify the conception of the atom as indestructible, immutable, impenetrable matter. It has parts, but it has no interstices of void; therefore no destroying agency can get between these parts and sever them. Hence we must recognize that, though the conception of atoms accounts for all composite bodies, analysis is not exhausted when these composite bodies have been reduced to atoms. There is a minimum smaller than the atom, but no such minimum separately exists. The atom is the least thing which can exist 'in solid singleness,' the limit of separate, individual existence. It would therefore be an error to suppose that minima of the atom exist at first apart and then combine to form atoms as atoms combine to form composite things. The minima of the atom are inseparable from each other and from the atom to all eternity.

The author is attempting to meet the objection that in infinite space there is no up and down, which he grants, if up and down are used in an absolute sense as implying a highest and a lowest point in infinite space. But he goes on to defend the use of the terms in a relative sense, and to deny that the same direction can be at once both up and down in reference to the same point of space. If it be granted that a line starting from a given point in a given direction may be produced both ways to infinity, then, he contends, if we call motion along this line in one direction up, we may also call motion along this line in the opposite direction down. A falling body which moves in the direction from our head to our feet and straight on in the same direction to infinity has for us a downward motion, and whatever moves in the contrary direction from our feet to our heads and straight on in the same direction to infinity has for us an upward motion. From the infinity of worlds it may be inferred that there are some worlds vertically over our heads and others beneath our feet; in the last sentence but one we seem to find a reference to the inhabitants of such worlds. A point on the vertical line may be 'down' from their stand-point, though it is 'up' from ours, or vice versa.

But Epicurus seems unconscious of the many assumptions which his statement involves. His atoms are absolutely hard and therefore inelastic. According to him the direction of motion changes after impact, but there is no loss of energy, and friction is ignored. His own concern is first with Democritus, whom apparently he charges with confounding motion in a medium such as air with motion in a void, and next with the interesting and different problem, to which we have already referred, of the motions of atoms in which looser or closer association form composite bodies. If we may expand the terse obscurity of the summary, the point he makes seems to be this. In motion of translation the whole composite body in finite time passes from point A to point B in a straight line. We are tempted, therefore, by the perversity of over-hasty presuppositions, and all those tendencies which we may call groundless opinion, inference, or belief, to argue that, if this finite time be subdivided into atoms of time distinctly conceivable by the mind but too short to be apprehended by sense, the uniform motion of translation will be maintained through each of them, not only for the composite moving body as a whole, but for each of its component atoms. This he brands as a mistake. We have clear and distinct apprehensions by the mind which are trustworthy, because in them the mind seizes and grasps objective images. When we picture the actual course of a single atom in a composite body moving with motion of translation, we see clearly and distinctly that it does not describe a free course, but is in perpetual oscillation backward and forward on account of collision with the other atoms associated with it in the composite body, and we may suppose him to add this is the gist of the argument, though nowhere expressed that in this perpetual oscillation backward and forward each atom of the composite body moves with uniform velocity 'quick as thought,' as if it were moving singly and freely through space, although the movement of translation of the whole composite body, as attested by sense, is so immeasurably slower than the motion of the atom.

Epicurus seems to have argued that vertical motion in the determinate direction which we call downward is prior to the motion resulting from collision, impact, and pressure, though why this should be so it is hard to see, and that atoms moving with equal velocity in the same direction would never collide. Feeling bound to offer some explanation, since both the tendency to fall downward and the collision seemed guaranteed by sense, he modified his premises in an arbitrary manner by the gratuitous assumption of an atomic declination from the perpendicular to a minimum extent. Sense tells us that heavy bodies fall downward to the earth, but sense never can assure us that they do not diverge from the perpendicular, provided the divergence is too small for sense to discern. Here, again, he avails himself of that convenient loose second clause of the canon with its fatal flaw: 'Nothing in our experience contradicts such an assumption.' Certainly not, when the assumption is expressly removed from the region of trustworthy observation. The all important evidence of sense does not, because it cannot contradict an imperceptible swerving.

Epicurus, like Democritus, supposed atoms moving in all directions, the inherent force of pseudo-gravity with which Epicurus, in obedience to experience, endowed his atoms, being everywhere counteracted by the effects of collision. The actual universe shows on a large scale what we see of motes in a sunbeam, viz., a dance of particles in all directions. The ceaseless rain of eternal atoms racing through infinite space in the same downward direction, the conception which called forth the enthusiasm of Fleeming Jenkin, belongs to an unreal or imaginary universe in which free atoms never collide because they never decline. Such a conception Epicurus relegated to the limbo of false opinion, unreality, and error for the sufficient reason that our world, and infinite other worlds, actually exist, i.e., have come into being, which could never have happened on the hypothesis rejected.

Atoms of soul can neither have sensation themselves nor cause the body to have sensation unless they are confined in the body. When so confined, they not only have sensation, but communicate it to the body, which becomes sentient. But other properties of the soul, e. g., the power to think, are not in this way communicated to the body, confinement in which is the indispensable condition that the soul should have sensation and thought. Special stress is laid on the mutual relation and inter-connection between the soul and the body, such that neither can exist without the other. We also learn that soul is a corporeal thing, a very fine substance, and a composite substance, wind and heat being mentioned as two elements in the compound.

Unlike empty space, which has real and separate existence, time is merely an accident, and, further, that to which it attaches, that of which it is an accident, is not anything real or corporeal but is itself an accident. Time, then, is an accident of accidents, an accident of events or occurrences in the present, past, or future. There is no plan in nature, says Epicurus, nothing which can be referred to supernatural will or agency.

Epicurus first points out that the intelligent government of the world is fatal to the immortality of bliss which is the divine prerogative, and then tenders a different explanation of the order and regularity of phenomena. The sun rises and sets regularly only because the combination of atoms evolves that particular change again and again with an approximation to uniformity.

When the same effect is known to have more than one cause, and we are uncertain to which of these causes it is to be referred in a particular case, then if we are sure that the question whether it is to be referred to cause A or to cause B does not affect our tranquillity, we need not carry the investigation any further. The knowledge that of all the causes which bring about this effect there is none that in any way disturbs our tranquillity, conduces to that tranquillity just as much as would the precise knowledge to which of these given causes the effect on a given occasion is due. How this principle works may be seen from the application made by Epicurus himself in the extant letter to Pythocles. In investigating a phenomenon of the class defined whose cause is unknown, Epicurus, on principle, stops short so soon as he has reached a plurality of causes any one of which is upon analogy judged capable of producing the effect under investigation without calling in supernatural agency. Over the results so obtained, which will appear to some ludicrous, to others lamentable, the friends of the philosopher will prefer to throw a veil.

In the letter to Herodotus Epicurus has given an epitome of his physical theory so adequate and yet so compressed that he recommends his pupil to commit it to memory. Once more, he emphasizes the subordination of all physical inquiries to ethical considerations. His sole aim is to banish for ever from the mind those fertile sources of disturbance, superstition, and terror. In so far as these anxieties are due to ignorance, their proper cure is knowledge, and within these bounds the pursuit of knowledge should be encouraged, not for its own sake far from it but as the indispensable means to the great end of life, the tranquillity of the individual.

The Writings of Epicurus

Principal Doctrines 🔖
(tr. Robert Drew Hicks)

1. A happy and eternal being has no trouble himself and brings no trouble upon any other being; hence he is exempt from movements of anger and partiality, for every such movement implies weakness.

2. Death is nothing to us; for the body, when it has been resolved into its elements, has no feeling, and that which has no feeling is nothing to us.

3. The magnitude of pleasure reaches its limit in the removal of all pain. When pleasure is present, so long as it is uninterrupted, there is no pain either of body or of mind or of both together.

4. Continuous pain does not last long in the body; on the contrary, pain, if extreme, is present a short time, and even that degree of pain which barely outweighs pleasure in the body does not last for many days together. Illnesses of long duration even permit of an excess of pleasure over pain in the body.

5. It is impossible to live a pleasant life without living wisely and well and justly, and it is impossible to live wisely and well and justly without living pleasantly. Whenever any one of these is lacking, when, for instance, the person is not able to live wisely, though he lives well and justly, it is impossible for him to live a pleasant life.

6. In order to obtain security from other people any means whatever of procuring this was a natural good.

7. Some people have sought to become famous and renowned, thinking that thus they would make themselves secure against their fellow-humans. If, then, the life of such persons really was secure, they attained natural good; if, however, it was insecure, they have not attained the end which by nature's own prompting they originally sought.

8. No pleasure is in itself evil, but the things which produce certain pleasures entail annoyances many times greater than the pleasures themselves.

9. If all pleasure had been capable of accumulation, – if this had gone on not only be recurrences in time, but all over the frame or, at any rate, over the principal parts of human nature, there would never have been any difference between one pleasure and another, as in fact there is.

10. If the objects which are productive of pleasures to profligate persons really freed them from fears of the mind, – the fears, I mean, inspired by celestial and atmospheric phenomena, the fear of death, the fear of pain; if, further, they taught them to limit their desires, we should never have any fault to find with such persons, for they would then be filled with pleasures to overflowing on all sides and would be exempt from all pain, whether of body or mind, that is, from all evil.

11. If we had never been molested by alarms at celestial and atmospheric phenomena, nor by the misgiving that death somehow affects us, nor by neglect of the proper limits of pains and desires, we should have had no need to study natural science.

12. It would be impossible to banish fear on matters of the highest importance, if a person did not know the nature of the whole universe, but lived in dread of what the legends tell us. Hence without the study of nature there was no enjoyment of unmixed pleasures.

13. There would be no advantage in providing security against our fellow humans, so long as we were alarmed by occurrences over our heads or beneath the earth or in general by whatever happens in the boundless universe.

14. When tolerable security against our fellow humans is attained, then on a basis of power sufficient to afford supports and of material prosperity arises in most genuine form the security of a quiet private life withdrawn from the multitude.

15. Nature's wealth at once has its bounds and is easy to procure; but the wealth of vain fancies recedes to an infinite distance.

16. Fortune but seldom interferes with the wise person; his greatest and highest interests have been, are, and will be, directed by reason throughout the course of his life.

17. The just person enjoys the greatest peace of mind, while the unjust is full of the utmost disquietude.

18. Pleasure in the body admits no increase when once the pain of want has been removed; after that it only admits of variation. The limit of pleasure in the mind, however, is reached when we reflect on the things themselves and their congeners which cause the mind the greatest alarms.

19. Unlimited time and limited time afford an equal amount of pleasure, if we measure the limits of that pleasure by reason.

20. The body receives as unlimited the limits of pleasure; and to provide it requires unlimited time. But the mind, grasping in thought what the end and limit of the body is, and banishing the terrors of futurity, procures a complete and perfect life, and has no longer any need of unlimited time. Nevertheless it does not shun pleasure, and even in the hour of death, when ushered out of existence by circumstances, the mind does not lack enjoyment of the best life.

21. He who understands the limits of life knows how easy it is to procure enough to remove the pain of want and make the whole of life complete and perfect.

Hence he has no longer any need of things which are not to be won save by labour and conflict.

22. We must take into account as the end all that really exists and all clear evidence of sense to which we refer our opinions; for otherwise everything will be full of uncertainty and confusion.

23. If you fight against all your sensations, you will have no standard to which to refer, and thus no means of judging even those judgments which you pronounce false.

24. If you reject absolutely any single sensation without stopping to discriminate with respect to that which awaits confirmation between matter of opinion and that which is already present, whether in sensation or in feelings or in any immediate perception of the mind, you will throw into confusion even the rest of your sensations by your groundless belief and so you will be rejecting the standard of truth altogether. If in your ideas based upon opinion you hastily affirm as true all that awaits confirmation as well as that which does not, you will not escape error, as you will be maintaining complete ambiguity whenever it is a case of judging between right and wrong opinion.

25. If you do not on every separate occasion refer each of your actions to the end prescribed by nature, but instead of this in the act of choice or avoidance swerve aside to some other end, your acts will not be consistent with your theories.

26. All such desires as lead to no pain when they remain ungratified are unnecessary, and the longing is easily got rid of, when the thing desired is difficult to procure or when the desires seem likely to produce harm.

27. Of all the means which are procured by wisdom to ensure happiness throughout the whole of life, by far the most important is the acquisition of friends.

28. The same conviction which inspires confidence that nothing we have to fear is eternal or even of long duration, also enables us to see that even in our limited conditions of life nothing enhances our security so much as friendship.

29. Of our desires some are natural and necessary others are natural, but not necessary; others, again, are neither natural nor necessary, but are due to illusory opinion.

30. Those natural desires which entail no pain when not gratified, though their objects are vehemently pursued, are also due to illusory opinion; and when they are not got rid of, it is not because of their own nature, but because of the person's illusory opinion.

31. Natural justice is a symbol or expression of usefulness, to prevent one person from harming or being harmed by another.

32. Those animals which are incapable of making covenants with one another, to the end that they may neither inflict nor suffer harm, are without either justice or injustice. And those tribes which either could not or would not form mutual covenants to the same end are in like case.

33. There never was an absolute justice, but only an agreement made in reciprocal association in whatever localities now and again from time to time, providing against the infliction or suffering of harm.

34. Injustice is not in itself an evil, but only in its consequence, viz. the terror which is excited by apprehension that those appointed to punish such offences will discover the injustice.

35. It is impossible for the person who secretly violates any article of the social compact to feel confident that he will remain undiscovered, even if he has already escaped ten thousand times; for right on to the end of his life he is never sure he will not be detected.

36. Taken generally, justice is the same for all, to wit, something found useful in mutual association; but in its application to particular cases of locality or conditions of whatever kind, it varies under different circumstances.

37. Among the things accounted just by conventional law, whatever in the needs of mutual association is attested to be useful, is thereby stamped as just, whether or not it be the same for all; and in case any law is made and does not prove suitable to the usefulness of mutual association, then this is no longer just. And should the usefulness which is expressed by the law vary and only for a time correspond with the prior conception, nevertheless for the time being it was just, so long as we do not trouble ourselves about empty words, but look simply at the facts.

38. Where without any change in circumstances the conventional laws, when judged by their consequences, were seen not to correspond with the notion of justice, such laws were not really just; but wherever the laws have ceased to be useful in consequence of a change in circumstances, in that case the laws were for the time being just when they were useful for the mutual association of the citizens, and subsequently ceased to be just when they ceased to be useful.

39. He who best knew how to meet fear of external foes made into one family all the creatures he could; and those he could not, he at any rate did not treat as aliens; and where he found even this impossible, he avoided all association, and, so far as was useful, kept them at a distance.

40. Those who were best able to provide themselves with the means of security against their neighbours, being thus in possession of the surest guarantee, passed the most agreeable life in each other's society; and their enjoyment of the fullest intimacy was such that, if one of them died before his time, the survivors did not mourn his death as if it called for sympathy.

LETTER TO HERODOTUS 🙐

(tr. Robert Drew Hicks, from *Lives of the Eminent Philosophers* by Diogenes Laertius)

Epicurus to Herodotus, greetings:

For those who are unable to study carefully all my physical writings or to go into the longer treatises at all, I have myself prepared an epitome of the whole system, Herodotus, to preserve in the memory enough of the principal doctrines, to the end that on every occasion they may be able to aid themselves on the most important points, so far as they take up the study of Physics. Those who have made some advance in the survey of the entire system ought to fix in their minds under the principal headings an elementary outline of the whole treatment of the subject. For a comprehensive view is often required, the details but seldom.

To the former, then – the main heads – we must continually return, and must memorize them so far as to get a valid conception of the facts, as well as the means of discovering all the details exactly when once the general outlines are rightly understood and remembered; since it is the privilege of the mature student to make a ready use of his conceptions by referring every one of them to elementary facts and simple terms. For it is impossible to gather up the results of continuous diligent study of the entirety of things, unless we can embrace in short formulas and hold in mind all that might have been accurately expressed even to the minutest detail.

Hence, since such a course is of service to all who take up natural science, I, who devote to the subject my continuous energy and reap the calm enjoyment of a life like this, have prepared for you just such an epitome and manual of the doctrines as a whole.

In the first place, Herodotus, you must understand what it is that words denote, in order that by reference to this we may be in a position to test opinions, inquiries, or problems, so that our proofs may not run on untested ad infinitum, nor the terms we use be empty of meaning. For the primary signification of every term employed must be clearly seen, and ought to need no proving; this being necessary, if we are to have something to which the point at issue or the problem or the opinion before us can be referred.

Next, we must by all means stick to our sensations, that is, simply to the present impressions whether of the mind or of any criterion whatever, and similarly to our actual feelings, in order that we may have the means of determining that which needs confirmation and that which is obscure.

When this is clearly understood, it is time to consider generally things which are obscure. To begin with, nothing comes into being out of what is non-existent. For in that case anything would have arisen out of anything, standing as it would in no need of its proper germs. And if that which disappears had been destroyed and become non-existent, everything would have perished, that into which the things were dissolved being non-existent. Moreover, the sum total of things was always such as it is now, and such it will ever remain. For there is nothing into which it can change. For outside the sum of things there is nothing which could enter into it and bring about the change.

Further, the whole of being consists of bodies and space. For the existence of bodies is everywhere attested by sense itself, and it is upon sensation that reason must rely when it attempts to infer the unknown from the known. And if there were no space (which we call also void and place and intangible nature), bodies would have nothing in which to be and through which to move, as they are plainly seen to move. Beyond bodies and space there is nothing which by mental apprehension or on its analogy we can conceive to exist. When we speak of bodies and space, both are regarded as wholes or separate things, not as the properties or accidents of separate things.

Again, of bodies some are composite, others the elements of which these composite bodies are made. These elements are indivisible and unchangeable, and necessarily so, if things are not all to be destroyed and pass into non-existence, but are to be strong enough to endure when the composite bodies are broken up, because they possess, a solid nature and are incapable of being anywhere or anyhow dissolved. It follows that the first beginnings must be indivisible, corporeal entities.

Again, the sum of things is infinite. For what is finite has an extremity, and the extremity of anything is discerned only by comparison with something else. Now the sum of things is not discerned by comparison with anything else: hence it has no extremity, it has no limit; and, since it has no limit, it must be unlimited or infinite.

Moreover, the sum of things is unlimited both by reason of the multitude of the atoms and the extent of the void. For if the void were infinite and bodies finite, the bodies would not have stayed anywhere but would have been dispersed in their course through the infinite void, not having any supports or counter-checks to send them back on their upward rebound. Again, if the void were finite, the infinity of bodies would not have anywhere to be.

Furthermore, the atoms, which have no void in them – out of which composite bodies arise and into which they are dissolved – vary indefinitely in their shapes; for so many varieties of things as we see could never have arisen out of a recurrence of a definite number of the same shapes. The like atoms of each shape are absolutely infinite; but the variety of shapes, though indefinitely large, is not absolutely infinite.

The atoms are in continual motion through all eternity. Some of them rebound to a considerable distance from each other, while others merely oscillate in one place when they chance to have got entangled or to be enclosed by a mass of other atoms shaped for entangling.

This is because each atom is separated from the rest by void, which is incapable of offering any resistance to the rebound; while it is the solidity of the atom which makes it rebound after a collision, however short the distance to which it rebounds, when it finds itself imprisoned in a mass of entangling atoms. Of all this there is no beginning, since both atoms and void exist from everlasting.

The repetition at such length of all that we are now recalling to mind furnishes an adequate outline for our conception of the nature of things.

Moreover, there is an infinite number of worlds, some like this world, others unlike it. For the atoms being infinite in number, as has just been proved, are borne ever further in their course. For the atoms out of which a world might arise, or by which a world might be formed, have not all been expended on one world or a finite number of worlds, whether like or unlike this one. Hence there will be nothing to hinder an infinity of worlds.

Again, there are outlines or films, which are of the same shape as solid bodies, but of a thinness far exceeding that of any object that we see. For it is not impossible that there should be found in the surrounding air combinations of this kind, materials adapted for expressing the hollowness and thinness of surfaces, and effluxes preserving the same relative position and motion which they had in the solid objects from which they come. To these films we give the name of 'images' or 'idols'. Furthermore, so long as nothing comes in the way to offer resistance, motion through the void accomplishes any imaginable distance in an inconceivably short time. For resistance encountered is the equivalent of slowness, its absence the equivalent of speed.

Not that, if we consider the minute times perceptible by reason alone, the moving body itself arrives at more than one place simultaneously (for this too is inconceivable), although in time perceptible to sense it does arrive simultaneously, however different the point of departure from that conceived by us. For if it changed its direction, that would be equivalent to its meeting with resistance, even if up to that point we allow nothing to impede the rate of its flight. This is an elementary fact which in itself is well worth bearing in mind. In the next place the exceeding thinness of the images is contradicted by none of the facts under our observation. Hence also their velocities are enormous, since they always find a void passage to fit them. Besides, their incessant effluence meets with no resistance or very little, although many atoms, not to say an unlimited number, do at once encounter resistance.

Besides this, remember that the production of the images is as quick as thought. For particles are continually streaming off from the surface of bodies, though no diminution of the bodies is observed, because other particles take their place. And those given off for a long time retain the position and arrangement which their atoms had when they formed part of the solid bodies, although occasionally they are thrown into confusion. Sometimes such films are formed very rapidly in the air, because they need not have any solid content; and there are other modes in which they may be formed. For there is nothing in all this which is contradicted by sensation, if we in some sort look at the clear evidence of sense, to which we should also refer the continuity of particles in the objects external to ourselves.

We must also consider that it is by the entrance of something coming from external objects that we see their shapes and think of them. For external things would not stamp on us their own nature of colour and form through the medium of the air which is between them and use or by means of rays of light or currents of any sort going from us to them, so well as by the entrance into our eyes or minds, to whichever their size is suitable, of certain films coming from the things themselves, these films or outlines being of the same color and shape as the external things themselves. They move with rapid motion; and this again explains why they present the appearance of the single continuous object, and retain the mutual interconnection which they had in the object, when they impinge upon the sense, such impact being due to the oscillation of the atoms in the interior of the solid object from which they come. And whatever presentation we derive by direct contact, whether it be with the mind or with the sense-organs, be it shape that is presented or other properties, this shape as presented is the shape of the solid thing, and it is due either to a close coherence of the image as a whole or to a mere remnant of its parts. Falsehood and error always depend upon the intrusion of opinion when a fact awaits confirmation or the absence of contradiction, which fact is afterwards frequently not confirmed or even contradicted following a certain movement in ourselves connected with, but distinct from, the mental picture presented – which is the cause of error.

For the presentations which, for example, are received in a picture or arise in dreams, or from any other form of apprehension by the mind or by the other criteria of truth, would never have resembled

what we call the real and true things, had it not been
for certain actual things of the kind with which we
come in contact. Error would not have occurred, if we
had not experienced some other movement in
ourselves, conjoined with, but distinct from, the
perception of what is presented. And from this
movement, if it be not confirmed or be contradicted,
falsehood results; while, if it be confirmed or not
contradicted, truth results.

And to this view we must closely adhere, if
we are not to repudiate the criteria founded on the
clear evidence of sense, nor again to throw all these
things into confusion by maintaining falsehood as
if it were truth.

Again, hearing takes place when a current passes
from the object, whether person or thing, which emits
voice or sound or noise, or produces the sensation of
hearing in any way whatever. This current is broken
up into homogeneous particles, which at the same time
preserve a certain mutual connection and a distinctive
unity extending to the object which emitted them, and
thus, for the most part, cause the perception in that
case or, if not, merely indicate the presence of the
external object. For without the transmission from the
object of a certain interconnection of the parts no such
sensation could arise. Therefore we must not suppose
that the air itself is molded into shape by the voice
emitted or something similar; for it is very far from
being the case that the air is acted upon by it in this
way. The blow which is struck in us when we utter a
sound causes such a displacement of the particles as
serves to produce a current resembling breath, and this

displacement gives rise to the sensation of hearing.

Again, we must believe that smelling, like hearing, would produce no sensation, were there not particles conveyed from the object which are of the proper sort for exciting the organ of smelling, some of one sort, some of another, some exciting it confusedly and strangely, others quietly and agreeably.

Moreover, we must hold that the atoms in fact possess none of the qualities belonging to things which come under our observation, except shape, weight, and size, and the properties necessarily conjoined with shape. For every quality changes, but the atoms do not change, since, when the composite bodies are dissolved, there must needs be a permanent something, solid and indissoluble, left behind, which makes change possible: not changes into or from the non-existent, but often through differences of arrangement, and sometimes through additions and subtractions of the atoms. Hence these somethings capable of being diversely arranged must be indestructible, exempt from change, but possessed each of its own distinctive mass and configuration. This must remain.

For in the case of changes of configuration within our experience the figure is supposed to be inherent when other qualities are stripped of, but the qualities are not supposed, like the shape which is left behind, to inhere in the subject of change, but to vanish altogether from the body. Thus, then, what is left behind is sufficient to account for the differences in composite bodies, since something at least must necessarily be left remaining and be immune from annihilation.

Again, you should not suppose that the atoms have any and every size, lest you be contradicted by facts; but differences of size must be admitted; for this addition renders the facts of feeling and sensation easier of explanation. But to attribute any and every magnitude to the atoms does not help to explain the differences of quality in things; moreover, in that case atoms large enough to be seen ought to have reached us, which is never observed to occur; nor can we conceive how its occurrence should be possible, in other words that an atom should become visible.

Besides, you must not suppose that there are parts unlimited in number, be they ever so small, in any finite body. Hence not only must we reject as impossible subdivision ad infinitum into smaller and smaller parts, lest we make all things too weak and, in our conceptions of the aggregates, be driven to pulverize the things that exist, in other words the atoms, and annihilate them; but in dealing with finite things we must also reject as impossible the progression ad infinitum by less and less increments.

For when once we have said that an infinite number of particles, however small, are contained in anything, it is not possible to conceive how it could any longer be limited or finite in size. For clearly our infinite number of particles must have some size; and then, of whatever size they were, the aggregate they made would be infinite. And, in the next place, since what is finite has an extremity which is distinguishable, even if it is not by itself observable, it is not possible to avoid thinking of another such extremity next to this. Nor can we help thinking that in this way, by proceeding forward from one to the next in order, it is possible by such a progression to arrive in thought at infinity.

We must consider the minimum perceptible by sense as not corresponding to that which is capable of being traversed, that is to say is extended, nor again as utterly unlike it, but as having something in common with the things capable of being traversed, though it is without distinction of parts. But when from the illusion created by this common property we think we shall distinguish something in the minimum, one part on one side and another part on the other side, it must be another minimum equal to the first which catches our eye. In fact, we see these minima one after another, beginning with the first, and not as occupying the same space; nor do we see them touch one another's parts with their parts, but we see that by virtue of their own peculiar character (as being unit indivisibles) they afford a means of measuring magnitudes: there are more of them, if the magnitude measured is greater; fewer of them, if the magnitude measured is less.

We must recognize that this analogy also holds of the minimum in the atom; it is only in minuteness that it differs from that which is observed by sense, but it follows the same analogy. On the analogy of things within our experience we have declared that the atom has magnitude; and this, small as it is, we have merely reproduced on a larger scale. And further, the least and simplest things must be regarded as extremities of lengths, furnishing from themselves as units the means of measuring lengths, whether greater or less, the mental vision being employed, since direct observation is impossible. For the community which exists between them and the unchangeable parts (the minimal parts of area or surface) is sufficient to justify the conclusion so far as this goes. But it is not possible that these minima of the atom should group themselves together through the possession of motion.

Further, we must not assert 'up' and 'down' of that which is unlimited, as if there were a zenith or nadir. As to the space overhead, however, if it be possible to draw a line to infinity from the point where we stand, we know that never will this space – or, for that matter, the space below the supposed standpoint if produced to infinity – appear to us to be at the same time 'up' and 'down' with reference to the same point; for this is inconceivable. Hence it is possible to assume one direction of motion, which we conceive as extending upwards ad infinitum, and another downwards, even if it should happen ten thousand times that what moves from us to the spaces above our heads reaches the feet of those above us, or that which moves downwards from us the heads of those below us. None the less is it true that the whole of the motion in the respective cases is conceived as extending in opposite directions ad infinitum.

When they are travelling through the void and meet with no resistance, the atoms must move with equal speed. Neither will heavy atoms travel more quickly than small and light ones, so long as nothing meets them, nor will small atoms travel more quickly than large ones, provided they always find a passage suitable to their size. and provided also that they meet with no obstruction. Nor will their upward or their lateral motion, which is due to collisions, nor again their downward motion, due to weight, affect their velocity. As long as either motion obtains, it must continue, quick as the speed of thought, provided there is no obstruction, whether due to external collision or to the atoms' own weight counteracting the force of the blow.

Moreover, when we come to deal with composite bodies, one of them will travel faster than another, although their atoms have equal speed. This is because the atoms in the aggregates are travelling in one direction during the shortest continuous time, albeit they move in different directions in times so short as to be appreciable only by the reason, but frequently collide until the continuity of their motion is appreciated by sense. For the assumption that beyond the range of direct observation even the minute times conceivable by reason will present continuity of motion is not true in the case before us. Our canon is that direct observation by sense and direct apprehension by the mind are alone invariably true.

Next, keeping in view our perceptions and feelings (for so shall we have the surest grounds for belief), we must recognize generally that the soul is a corporeal thing, composed of fine particles, dispersed all over the frame, most nearly resembling wind with an admixture of heat, in some respects like wind, in others like heat. But, again, there is the third part which exceeds the other two in the fineness of its particles and thereby keeps in closer touch with the rest of the frame. And this is shown by the mental faculties and feelings, by the ease with which the mind moves, and by thoughts, and by all those things the loss of which causes death. Further, we must keep in mind that soul has the greatest share in causing sensation. Still, it would not have had sensation, had it not been somehow confined within the rest of the frame. But the rest of the frame, though it provides this indispensable condition for the soul, itself also has a share, derived from the soul, of the said quality; and yet does not possess all the qualities of soul. Hence on the departure of the soul it loses sentience. For it had not this power in itself; but something else, congenital with the body, supplied it to body: which other thing, through the potentiality actualized in it by means of motion, at once acquired for itself a quality of sentience, and, in virtue of the neighbourhood and interconnection between them, imparted it (as I said) to the body also.

Hence, so long as the soul is in the body, it never loses sentience through the removal of some other part. The containing sheaths may be dislocated in whole or in part, and portions of the soul may thereby be lost; yet in spite of this the soul, if it manage to survive, will have sentience. But the rest of the frame, whether the whole of it survives or only a part, no longer has sensation, when once those atoms have departed, which, however few in number, are required to constitute the nature of soul. Moreover, when the whole frame is broken up, the soul is scattered and has no longer the same powers as before, nor the same notions; hence it does not possess sentience either.

For we cannot think of it as sentient, except it be in this composite whole and moving with these movements; nor can we so think of it when the sheaths which enclose and surround it are not the same as those in which the soul is now located and in which it performs these movements.

There is the further point to be considered, what the incorporeal can be, if, I mean, according to current usage the term is applied to what can be conceived as self-existent. But it is impossible to conceive anything that is incorporeal as self-existent except empty space. And empty space cannot itself either act or be acted upon, but simply allows body to move through it. Hence those who call soul incorporeal speak foolishly. For if it were so, it could neither act nor be acted upon. But, as it is, both these properties, you see, plainly belong to soul.

If, then, we bring all these arguments concerning soul to the criterion of our feelings and perceptions, and if we keep in mind the proposition stated at the outset, we shall see that the subject has been adequately comprehended in outline: which will enable us to determine the details with accuracy and confidence.

Moreover, shapes and colours, magnitudes and weights, and in short all those qualities which are predicated of body, in so far as they are perpetual properties either of all bodies or of visible bodies, are knowable by sensation of these very properties: these, I say, must not be supposed to exist independently by themselves (for that is inconceivable), nor yet to be non-existent, nor to be some other and incorporeal entities cleaving to body, nor again to be parts of body. We must consider the whole body in a general way to derive its permanent nature from all of them, though it is not, as it were, formed by grouping them together in the same way as when from the particles themselves a larger aggregate is made up, whether these particles be primary or any magnitudes whatsoever less than the particular whole. All these qualities, I repeat, merely give the body its own permanent nature. They all have their own characteristic modes of being perceived and distinguished, but always along with the whole body in which they inhere and never in separation from it; and it is in virtue of this complete conception of the body as a whole that it is so designated.

Again, qualities often attach to bodies without being permanent concomitants. They are not to be classed among invisible entities nor are they incorporeal. Hence, using the term 'accidents' in the commonest sense, we say plainly that 'accidents' have not the nature of the whole thing to which they belong, and to which, conceiving it as a whole, we give the name of body, nor that of the permanent properties without which body cannot be thought of. And in virtue of certain peculiar modes of apprehension into which the complete body always enters, each of them can be called an accident. But only as often as they are seen actually to belong to it, since such accidents are not perpetual concomitants. There is no need to banish from reality this clear evidence that the accident has not the nature of that whole – by us called body – to which it belongs, nor of the permanent properties which accompany the whole. Nor, on the other hand, must we suppose the accident to have independent existence (for this is just as inconceivable in the case of accidents as in that of the permanent properties); but, as is manifest, they should all be regarded as accidents, not as permanent concomitants, of bodies, nor yet as having the rank of independent existence. Rather they are seen to be exactly as and what sensation itself makes them individually claim to be.

There is another thing which we must consider carefully. We must not investigate time as we do the other accidents which we investigate in a subject, namely, by referring them to the preconceptions envisaged in our minds; but we must take into account the plain fact itself, in virtue of which we speak of time as long or short, linking to it in intimate connection this attribute of duration. We need not adopt any fresh terms as preferable, but should employ the usual expressions about it. Nor need we predicate anything else of time, as if this something else contained the same essence as is contained in the proper meaning of the word 'time' (for this also is done by some). We must chiefly reflect upon that to which we attach this peculiar character of time, and by which we measure it. No further proof is required: we have only to reflect that we attach the attribute of time to days and nights and their parts, and likewise to feelings of pleasure and pain and to neutral states, to states of movement and states of rest, conceiving a peculiar accident of these to be this very characteristic which we express by the word 'time'.

After the foregoing we have next to consider that the worlds and every finite aggregate which bears a strong resemblance to things we commonly see have arisen out of the infinite. For all these, whether small or great, have been separated off from special conglomerations of atoms; and all things are again dissolved, some faster, some slower, some through the action of one set of causes, others through the action of another.

And further, we must not suppose that the worlds have necessarily one and the same shape. For nobody can prove that in one sort of world there might not be contained, whereas in another sort of world there could not possibly be, the seeds out of which animals and plants arise and all the rest of the things we see.

Again, we must suppose that nature too has been taught and forced to learn many various lessons by the facts themselves, that reason subsequently develops what it has thus received and makes fresh discoveries, among some tribes more quickly, among others more slowly, the progress thus made being at certain times and seasons greater, at others less.

Hence even the names of things were not originally due to convention, but in the several tribes under the impulse of special feelings and special presentations of sense primitive man uttered special cries. The air thus emitted was moulded by their individual feelings or sense-presentations, and differently according to the difference of the regions which the tribes inhabited. Subsequently whole tribes adopted their own special names, in order that their communications might be less ambiguous to each other and more briefly expressed. And as for things not visible, so far as those who were conscious of them tried to introduce any such notion, they put in circulation certain names for them, either sounds which they were instinctively compelled to utter or which they selected by reason on analogy according to the most general cause there can be for expressing oneself in such a way.

Nay more: we are bound to believe that in the sky revolutions, solstices, eclipses, risings and settings, and the like, take place without the ministration or command, either now or in the future, of any being who at the same time enjoys perfect bliss along with immortality. For troubles and anxieties and feelings of anger and partiality do not accord with bliss, but always imply weakness and fear and dependence upon one's neighbours. Nor, again, must we hold that things which are no more than globular masses of fire, being at the same time endowed with bliss, assume these motions at will. Nay, in every term we use we must hold fast to all the majesty which attaches to such notions as bliss and immortality, lest the terms should generate opinions inconsistent with this majesty. Otherwise such inconsistency will of itself suffice to produce the worst disturbance in our minds. Hence, where we find phenomena invariably recurring, the invariability of the recurrence must be ascribed to the original interception and conglomeration of atoms whereby the world was formed.

Further, we must hold that to arrive at accurate knowledge of the cause of things of most moment is the business of natural science, and that happiness depends on this (viz. on the knowledge of celestial and atmospheric phenomena), and upon knowing what the heavenly bodies really are, and any kindred facts contributing to exact knowledge in this respect.

Further, we must recognize on such points as this

no plurality of causes or contingency, but must hold that nothing suggestive of conflict or disquiet is compatible with an immortal and blessed nature. And the mind can grasp the absolute truth of this.

But when we come to subjects for special inquiry, there is nothing in the knowledge of risings and settings and solstices and eclipses and all kindred subjects that contributes to our happiness; but those who are well-informed about such matters and yet are ignorant – what the heavenly bodies really are, and what are the most important causes of phenomena, feel quite as much fear as those who have no such special information – nay, perhaps even greater fear, when the curiosity excited by this additional knowledge cannot find a solution or understand the subordination of these phenomena to the highest causes.

Hence, if we discover more than one cause that may account for solstices, settings and risings, eclipses and the like, as we did also in particular matters of detail, we must not suppose that our treatment of these matters fails of accuracy, so far as it is needful to ensure our tranquillity and happiness. When, therefore, we investigate the causes of celestial and atmospheric phenomena, as of all that is unknown, we must take into account the variety of ways in which analogous occurrences happen within our experience; while as for those who do not recognize the difference between what is or comes about from a single cause and that which may be the effect of any one of several causes, overlooking the fact that the objects are only seen at a distance, and are moreover ignorant of the conditions

that render, or do not render, peace of mind impossible
– all such persons we must treat with contempt. If then
we think that an event could happen in one or other
particular way out of several, we shall be as tranquil
when we recognize that it actually comes about in
more ways than one as if we knew that it happens in
this particular way.

There is yet one more point to seize, namely, that the
greatest anxiety of the human mind arises through the
belief that the heavenly bodies are blessed and
indestructible, and that at the same time they have
volition and actions and causality inconsistent with this
belief; and through expecting or apprehending some
ever-lasting evil, either because of the myths, or because
we are in dread of the mere insensibility of death, as if it
had to do with us; and through being reduced to this
state not by conviction but by a certain irrational
perversity, so that, if men do not set bounds to their
terror, they endure as much or even more intense
anxiety than the man whose views on these matters are
quite vague. But mental tranquillity means being
released from all these troubles and cherishing a
continual remembrance of the highest and most
important truths.

Hence we must attend to present feelings and
sense perceptions, whether those of mankind in
general or those peculiar to the individual, and also
attend to all the clear evidence available, as given by
each of the standards of truth. For by studying them
we shall rightly trace to its cause and banish the
source of disturbance and dread, accounting for
celestial phenomena and for all other things which

from time to time befall us and cause the utmost alarm to the rest of mankind.

Here then, Herodotus, you have the chief doctrines of Physics in the form of a summary. So that, if this statement be accurately retained and take effect, a man will, I make no doubt, be incomparably better equipped than his fellows, even if he should never go into all the exact details. For he will clear up for himself many of the points which I have worked out in detail in my complete exposition; and the summary itself, if borne in mind, will be of constant service to him.

It is of such a sort that those who are already tolerably, or even perfectly, well acquainted with the details can, by analysis of what they know into such elementary perceptions as these, best prosecute their researches in physical science as a whole; while those, on the other hand, who are not altogether entitled to rank as mature students can in silent fashion and as quick as thought run over the doctrines most important for their peace of mind.

LETTER TO MENOECEUS ✎
(tr. Robert Drew Hicks from *Lives of the Eminent Philosophers* by Diogenes Laertius)

Greeting.

Let no one be slow to seek wisdom when he is young nor weary in the search thereof when he is grown old. For no age is too early or too late for the health of the soul. And to say that the season for studying philosophy has not yet come, or that it is past and gone, is like saying that the season for happiness is not yet or that it is now no more. Therefore, both old and young ought to seek wisdom, the former in order that, as age comes over him, he may be young in good things because of the grace of what has been, and the latter in order that, while he is young, he may at the same time be old, because he has no fear of the things which are to come. So we must exercise ourselves in the things which bring happiness, since, if that be present, we have everything, and, if that be absent, all our actions are directed toward attaining it.

  Those things which without ceasing I have declared to you, those do, and exercise yourself in those, holding them to be the elements of right life. First believe that God is a living being immortal and happy, according to the notion of a god indicated by the common sense of humankind; and so of him anything that is at agrees not with about him whatever may uphold both his happiness and his immortality. For truly there are gods, and knowledge of them is evident; but they are not such as the multitude believe, seeing that people do not steadfastly maintain the notions they form respecting them. Not the person who denies the gods worshipped by the multitude, but he who affirms of the gods what the multitude believes about them is truly impious. For the utterances of the multitude about the gods are not true preconceptions but false assumptions; hence it is that the greatest evils happen to the wicked and the greatest blessings happen to the good from the hand of the gods, seeing that they are always favourable to their own good qualities and take pleasure in people like to themselves, but reject as alien whatever is not of their kind.

Accustom yourself to believe that death is nothing to us, for good and evil imply awareness, and death is the privation of all awareness; therefore a right understanding that death is nothing to us makes the mortality of life enjoyable, not by adding to life an unlimited time, but by taking away the yearning after immortality. For life has no terror; for those who thoroughly apprehend that there are no terrors for them in ceasing to live. Foolish, therefore, is the person who says that he fears death, not because it will pain when it comes, but because it pains in the prospect. Whatever causes no annoyance when it is present, causes only a groundless pain in the expectation. Death, therefore, the most awful of evils, is nothing to us, seeing that, when we are, death is not come, and, when death is come, we are not. It is nothing, then, either to the living or to the dead, for with the living it is not and the dead exist no longer. But in the world, at one time people shun death as the greatest of all evils, and at another time choose it as a respite from the evils in life. The wise person does not deprecate life nor does he fear the cessation of life. The thought of life is no offence to him, nor is the cessation of life regarded as an evil. And even as people choose of food not merely and simply the larger portion, but the more pleasant, so the wise seek to enjoy the time which is most pleasant and not merely that which is longest. And he who admonishes the young to live well and the old to make a good end speaks foolishly, not merely because of the desirability of life, but because the same exercise at once teaches to live well and to die well. Much worse is he who says that it were good not to be born, but when once one is born to pass with all speed through the gates of Hades. For if he truly believes this, why does he not depart from life? It were easy for him to do so, if once he were firmly convinced. If he speaks only in mockery, his words are foolishness, for those who hear believe him not.

We must remember that the future is neither wholly ours nor wholly not ours, so that neither must we count upon it as quite certain to come nor despair of it as quite certain not to come.

We must also reflect that of desires some are natural, others are groundless; and that of the natural some are necessary as well as natural, and some natural only. And of the necessary desires some are necessary if we are to be happy, some if the body is to be rid of uneasiness, some if we are even to live. He who has a clear and certain understanding of these things will direct every preference and aversion toward securing health of body and tranquillity of mind, seeing that this is the sum and end of a happy life. For the end of all our actions is to be free from pain and fear, and, when once we have attained all this, the tempest of the soul is laid; seeing that the living creature has no need to go in search of something that is lacking, nor to look anything else by which the good of the soul and of the body will be fulfilled. When we are pained pleasure, then, and then only, do we feel the need of pleasure. For this reason we call pleasure the alpha and omega of a happy life. Pleasure is our first and kindred good. It is the starting-point of every choice and of every aversion, and to it we come back, inasmuch as we make feeling the rule by which to judge of every good thing. And since pleasure is our first and native good, for that reason we do not choose every pleasure whatever, but often pass over many pleasures when a greater annoyance ensues from them. And often we consider pains superior to pleasures when submission to the pains for a long time brings us as a consequence a greater pleasure. While therefore all pleasure because it is naturally akin to us is good, not all pleasure is worthy of choice, just as all pain is an evil and yet not all pain is to be shunned. It is, however, by measuring one against another, and by looking at the conveniences and inconveniences, that all these matters must be judged. Sometimes we treat the good as an evil, and the evil, on the contrary, as a good. Again, we regard independence of outward things as a great good, not so as in all cases to use little, but so as to be contented with little if we have not much, being honestly persuaded that they have the sweetest enjoyment of luxury who stand least in need of it, and that whatever is natural is easily procured and only the vain and worthless hard to win. Plain fare gives as much pleasure as a costly diet, when once the pain of want has been removed, while bread and water confer the highest possible pleasure when they are brought to hungry lips. To habituate one's self, therefore, to simple and inexpensive diet supplies all that is needful for health, and enables a person to meet the necessary requirements of life without shrinking and it places us in a better condition when we approach at intervals a costly fare and renders us fearless of fortune.

When we say, then, that pleasure is the end and aim, we do not mean the pleasures of the prodigal or the pleasures of sensuality, as we are understood to do by some through ignorance, prejudice, or willful misrepresentation. By pleasure we mean the absence of pain in the body and of trouble in the soul. It is not an unbroken succession of drinking-bouts and of merrymaking, not sexual love, not the enjoyment of the fish and other delicacies of a luxurious table, which produce a pleasant life; it is sober reasoning, searching out the grounds of every choice and avoidance, and banishing those beliefs through which the greatest disturbances take possession of the soul. Of all this the end is prudence. For this reason prudence is a more precious thing with which you will live as a god among people. For people lose all appearance of mortality by living in the midst of immortal blessings.

LETTER TO PYTHOCLES 🖎
(tr. Robert Drew Hicks from *Lives of the Eminent Philosophers* by Diogenes Laertius)

Epicurus to Pythocles, greeting:

In your letter to me, of which Cleon was the bearer, you continue to show me affection which I have merited by my devotion to you, and you try, not without success, to recall the considerations which make for a happy life. To aid your memory you ask me for a clear and concise statement respecting celestial phenomena; for what we have written on this subject elsewhere is, you tell me, hard to remember, although you have my books constantly with you. I was glad to receive your request and am full of pleasant expectations. We will then complete our writing and grant all you ask. Many others besides you will find these reasonings useful, and especially those who have but recently made acquaintance with the true story of nature and those who are attached to pursuits which go deeper than any part of ordinary education. So you will do well to take and learn them and get them up quickly along with the short epitome in my letter to Herodotus.

In the first place, remember that, like everything else, knowledge of celestial phenomena, whether taken along with other things or in isolation, has no other end in view than peace of mind and firm convictions. We do not seek to wrest by force what is impossible, nor to understand all matters equally well, nor make our treatment always as clear as when we discuss human life or explain the principles of physics in general – for instance, that the whole of being consists of bodies and intangible nature, or that the ultimate elements of things are indivisible, or any other proposition which admits only one explanation of the phenomena to be possible. But this is not the case with celestial phenomena: these at any rate admit of manifold causes for their occurrence and manifold accounts, none of them contradictory of sensation, of their nature.

For in the study of nature we must not conform to empty assumptions and arbitrary laws, but follow the promptings of the facts; for our life has no need now of unreason and false opinion; our one need is untroubled existence. All things go on uninterruptedly, if all be explained by the method of plurality of causes in conformity with the facts, so soon as we duly understand what may be plausibly alleged respecting them. But when we pick and choose among them, rejecting one equally consistent with the phenomena, we clearly fall away from the study of nature altogether and tumble into myth. Some phenomena within our experience afford evidence by which we may interpret what goes on in the heavens. We see how the former really take place, but not how the celestial phenomena take place, for their occurrence may possibly be due to a variety of causes. However, we must observe each fact as presented, and further separate from it all the facts presented along with it, the occurrence of which from various causes is not contradicted by facts within our experience.

A world is a circumscribed portion of the universe,

which contains stars and earth and all other visible things, cut off from the infinite, and terminating in an exterior which may either revolve or be at rest, and be round or triangular or of any other shape whatever. All these alternatives are possible: they are contradicted by none of the facts in this world, in which an extremity can nowhere be discerned.

That there is an infinite number of such worlds can be perceived, and that such a world may arise in a world or in one of the intermundia (by which term we mean the spaces between worlds) in a tolerably empty space and not, as some maintain, in a vast space perfectly clear and void. It arises when certain suitable seeds rush in from a single world or intermundium, or from several, and undergo gradual additions or articulations or changes of place, it may be, and waterings from appropriate sources, until they are matured and firmly settled in so far as the foundations laid can receive them. For it is not enough that there should be an aggregation or a vortex in the empty space in which a world may arise, as the necessitarians hold, and may grow until it collide with another, as one of the so-called physicists says. For this is in conflict with facts.

The sun and moon and the stars generally were not of independent origin and later absorbed, within our world, [such parts of it at least as serve at all for its defence]; but they at once began to take form and grow [and so too did earth and sea] by the accretions and whirling motions of certain substances of finest texture, of the nature either of wind or fire, or of both; for thus sense itself suggests.

The size of the sun and the remaining stars relatively to us is just as great as it appears. But in itself and actually it may be a little larger or a little smaller, or precisely as great as it is seen to be. For so too fires of which we have experience are seen by sense when we see them at a distance. And every objection brought against this part of the theory will easily be met by anyone who attends to plain facts, as I show in my work *On Nature*. And the rising and setting of the sun, moon, and stars may be due to kindling and quenching, provided that the circumstances are such as to produce this result in each of the two regions, east and west: for no fact testifies against this. Or the result might be produced by their coming forward above the earth and again by its intervention to hide them: for no fact testifies against this either. And their motions may be due to the rotation of the whole heaven, or the heaven may be at rest and they alone rotate according to some necessary impulse to rise, implanted at first when the world was made ... and this through excessive heat, due to a certain extension of the fire which always encroaches upon that which is near it.

The turnings of the sun and moon in their course may be due to the obliquity of the heaven, whereby it is forced back at these times. Again, they may equally be due to the contrary pressure of the air or, it may be, to the fact that either the fuel from time to time necessary has been consumed in the vicinity or there is a dearth of it. Or even because such a whirling motion was from the first inherent in these stars so that they move in a sort of spiral. For all such explanations and the like do not conflict with any clear evidence, if only in such details we hold fast to what is possible, and can bring each of these explanations into accord with the facts, unmoved by the servile artifices of the astronomers.

The waning of the moon and again her waxing might be due to the rotation of the moon's body, and equally well to configurations which the air assumes; further, it may be due to the interposition of certain bodies. In short, it may happen in any of the ways in which the facts within our experience suggest such an appearance to be explicable. But one must not be so much in love with the explanation by a single way as wrongly to reject all the others from ignorance of what can, and what cannot, be within human knowledge, and consequent longing to discover the undiscoverable. Further, the moon may possibly shine by her own light, just as possibly she may derive her light from the sun; for in our own experience we see many things which shine by their own light and many also which shine by borrowed light. And none of the celestial phenomena stand in the way, if only we always keep in mind the method of plural explanation and the several consistent assumptions and causes, instead of dwelling on what is inconsistent and giving it a false importance so as always to fall back in one way or another upon the single explanation. The appearance of the face in the moon may equally well arise from interchange of parts, or from interposition of something, or in any other of the ways which might be seen to accord with the facts. For in all the celestial phenomena such a line of research is not to be abandoned; for, if you fight against clear evidence, you never can enjoy genuine peace of mind.

An eclipse of the sun or moon may be due to the extinction of their light, just as within our own experience this is observed to happen; and again by interposition of something else – whether it be the earth or some other invisible body like it. And thus we must take in conjunction the explanations which agree with one another, and remember that the concurrence of more than one at the same time may not impossibly happen.

And further, let the regularity of their orbits be explained in the same way as certain ordinary incidents within our own experience; the divine nature must not on any account be adduced to explain this, but must be kept free from the task and in perfect bliss. Unless this be done, the whole study of celestial phenomena will be in vain, as indeed it has proved to be with some who did not lay hold of a possible method, but fell into the folly of supposing that these events happen in one single way only and of rejecting all the others which are possible, suffering themselves to be carried into the realm of the unintelligible, and being unable to take a comprehensive view of the facts which must be taken as clues to the rest.

The variations in the length of nights and days may be due to the swiftness and again to the slowness of the sun's motion in the sky, owing to the variations in the length of spaces traversed and to his accomplishing some distances more swiftly or more slowly, as happens sometimes within our own experience; and with these facts our explanation of celestial phenomena must agree; whereas those who adopt only one explanation are in conflict with the facts and are utterly mistaken as to the way in which man can attain knowledge.

The signs in the sky which betoken the weather may be due to mere coincidence of the seasons, as is the case with signs from animals seen on earth, or they may be caused by changes and alterations in the air. For neither the one explanation nor the other is in conflict with facts, and it is not easy to see in which cases the effect is due to one cause or to the other.

Clouds may form and gather either because the air is condensed under the pressure of winds, or because atoms which hold together and are suitable to produce this result become mutually entangled, or because currents collect from the earth and the waters; and there are several other ways in which it is not impossible for the aggregations of such bodies into clouds to be brought about. And that being so, rain may be produced from them sometimes by their compression, sometimes by their transformation; or again may be caused by exhalations of moisture rising from suitable places through the air, while a more violent inundation is due to certain accumulations suitable for such discharge. Thunder may be due to the rolling of wind in the hollow parts of the clouds, as it is sometimes imprisoned in vessels which we use; or to the roaring of fire in them when blown by a wind, or to the rending and disruption of clouds, or to the friction and splitting up of clouds when they have become as firm as ice.

As in the whole survey, so in this particular point, the facts invite us to give a plurality of explanations.

Lightning too happens in a variety of ways. For when the clouds rub against each other and collide, that collocation of atoms which is the cause of fire generates lightning; or it may be due to the flashing forth from the clouds, by reason of winds, of particles capable of producing this brightness; or else it is squeezed out of the clouds when they have been condensed either by their own action or by that of the winds; or again, the light diffused from the stars may be enclosed in the clouds, then driven about by their motion and by that of the winds, and finally make its escape from the clouds; or light of the finest texture may be filtered through the clouds (whereby the clouds may be set on fire and thunder produced), and the motion of this light may make lightning; or it may arise from the combustion of wind brought about by the violence of its motion and the intensity of its compression; or, when the clouds are rent asunder by winds, and the atoms which generate fire are expelled, these likewise cause lightning to appear.

And it may easily be seen that its occurrence is possible in many other ways, so long as we hold fast to facts and take a general view of what is analogous to them. Lightning precedes thunder, when the clouds are constituted as mentioned above and the configuration which produces lightning is expelled at the moment when the wind falls upon the cloud, and the wind being rolled up afterwards produces the roar of thunder; or, if both are simultaneous, the lightning moves with a greater velocity towards it and the thunder lags behind, exactly as when persons who are striking blows are observed from a distance.

A thunderbolt is caused when winds are repeatedly collected, imprisoned, and violently ignited; or when a part is torn asunder and is more violently expelled downwards, the rending being due to the fact that the compression of the clouds has made the neighbouring parts more dense; or again it may be due like thunder merely to the expulsion of the imprisoned fire, when this has accumulated and been more violently inflated with wind and has torn the cloud, being unable to withdraw to the adjacent parts because it is continually more and more closely compressed [generally by some high mountain where thunderbolts mostly fall]. And there are several other ways in which thunderbolts may possibly be produced. Exclusion of myth is the sole condition necessary; and it will be excluded, if one properly attends to the facts and hence draws inferences to interpret what is obscure.

Fiery whirlwinds are due to the descent of a cloud forced downwards like a pillar by the wind in full force and carried by a gale round and round, while at the same time the outside wind gives the cloud a lateral thrust; or it may be due to a change of the wind which veers to all points of the compass as a current of air from above helps to force it to move; or it may be that a strong eddy of winds has been started and is unable to burst through laterally because the air around is closely condensed. And when they descend upon land, they cause what are called tornadoes, in accordance with the various ways in which they are produced through the force of the wind; and when let down upon the sea, they cause waterspouts.

Earthquakes may be due to the imprisonment of wind underground, and to its being interspersed with small masses of earth and then set in continuous motion, thus causing the earth to tremble. And the earth either takes in this wind from without or from the falling in of foundations, when undermined, into subterranean caverns, thus raising a wind in the imprisoned air. Or they may be due to the propagation of movement arising from the fall of many foundations and to its being again checked when it encounters the more solid resistance of earth. And there are many other causes to which these oscillations of the earth may be due.

Winds arise from time to time when foreign matter continually and gradually finds its way into the air; also through the gathering of great stores of water. The rest of the winds arise when a few of them fall into the many hollows and they are thus divided and multiplied.

Hail is caused by the firmer congelation and complete transformation, and subsequent distribution into drops, of certain particles resembling wind: also by the slighter congelation of certain particles of moisture and the vicinity of certain particles of wind which at one and the same time forces them together and makes them burst, so that they become frozen in parts and in the whole mass. The round shape of hailstones is not impossibly due to the extremities on all sides being melted and to the fact that, as explained, particles either of moisture or of wind surround them evenly on all sides and in every quarter, when they freeze.

Snow may be formed when a fine rain issues from the clouds because the pores are symmetrical and because of the continuous and violent pressure of the winds upon clouds which are suitable; and then this rain has been frozen on its way because of some violent change to coldness in the regions below the clouds. Or again, by congelation in clouds which have uniform density a fall of snow might occur through the clouds which contain moisture being densely packed in close proximity to each other; and these clouds produce a sort of compression and cause hail, and this happens mostly in spring. And when frozen clouds rub against each other, this accumulation of snow might be thrown off. And there are other ways in which snow might be formed.

Dew is formed when such particles as are capable of producing this sort of moisture meet each other from the air: again by their rising from moist and damp places, the sort of place where dew is chiefly formed, and their subsequent coalescence, so as to create moisture and fall downwards, just as in several cases something similar is observed to take place under our eyes. And the formation of hoar-frost is not different from that of dew, certain particles of such a nature becoming in some such way congealed owing to a certain condition of cold air.

Ice is formed by the expulsion from the water of the circular, and the compression of the scalene and acute-angled atoms contained in it; further by the accretion of such atoms from without, which being driven together cause the water to solidify after the expulsion of a certain number of round atoms.

The rainbow arises when the sun shines upon humid air; or again by a certain peculiar blending of light with air, which will cause either all the distinctive qualities of these colours or else some of them belonging to a single kind, and from the reflection of this light the air all around will be coloured as we see it to be, as the sun shines upon its parts. The circular shape which it assumes is due to the fact that the distance of every point is perceived by our sight to be equal; or it may be because, the atoms in the air or in the clouds and deriving from the sun having been thus united, the aggregate of them presents a sort of roundness.

A halo round the moon arises because the air on all sides extends to the moon; or because it equably raises upwards the currents from the moon so high as to impress a circle upon the cloudy mass and not to separate it altogether; or because it raises the air which immediately surrounds the moon symmetrically from all sides up to a circumference round her and there forms a thick ring. And this happens at certain parts either because a current has forced its way in from without or because the heat has gained possession of certain passages in order to effect this.

Comets arise either because fire is nourished in certain places at certain intervals in the heavens, if circumstances are favourable; or because at times the heaven has a particular motion above us so that such stars appear; or because the stars themselves are set in motion under certain conditions and come to our neighbourhood and show themselves. And their disappearance is due to the causes which are the opposite of these. Certain stars may revolve without setting not only for the reason alleged by some, because this is the part of the world round which, itself unmoved, the rest revolves, but it may also be because a circular eddy of air surrounds this part, which prevents them from travelling out of sight like other stars or because there is a dearth of necessary fuel farther on, while there is abundance in that part where they are seen to be. Moreover there are several other ways in which this might be brought about, as may be seen by anyone capable of reasoning in accordance with the facts.

The wanderings of certain stars, if such wandering is their actual motion, and the regular movement of certain other stars, may be accounted for by saying that they originally moved in a circle and were constrained, some of them to be whirled round with the same uniform rotation and others with a whirling motion which varied; but it may also be that according to the diversity of the regions traversed in some places there are uniform tracts of air, forcing them forward in one direction and burning uniformly, in others these tracts present such irregularities as cause the motions observed. To assign a single cause for these effects when the facts suggest several causes is madness and a strange inconsistency; yet it is done by adherents of rash astronomy, who assign meaningless causes for the stars whenever they persist in saddling the divinity with burdensome tasks. That certain stars are seen to be left behind by others may be because they travel more slowly, though they go the same round as the others; or it may be that they are drawn back by the same whirling motion and move in the opposite direction; or again it may be that some travel over a larger and others over a smaller space in making the same revolution. But to lay down as assured a single explanation of these phenomena is worthy of those who seek to dazzle the multitude with marvels.

Falling stars, as they are called, may in some cases be due to the mutual friction of the stars themselves, in other cases to the expulsion of certain parts when that mixture of fire and air takes place which was mentioned when we were discussing lightning; or it may be due to the meeting of atoms capable of generating fire, which accord so well as to produce this result, and their subsequent motion wherever the impulse which brought them together at first leads them; or it may be that wind collects in certain dense mist-like masses and, since it is imprisoned, ignites and then bursts forth upon whatever is round about it, and is carried to that place to which its motion impels it. And there are other ways in which this can be brought about without recourse to myths.

The fact that the weather is sometimes foretold from the behaviour of certain animals is a mere coincidence in time. For the animals offer no necessary reason why a storm should be produced and no divine being sits observing when these animals go out and afterwards fulfilling the signs which they have given. For such folly as this would not possess the most ordinary being if ever so little enlightened, much less one who enjoys perfect felicity.

All this, Pythocles, you should keep in mind; for then you will escape a long way from myth, and you will be able to view in their connection the instances which are similar to these. But above all give yourself up to the study of first principles and of infinity and of kindred subjects, and further of the standards and of the feelings and of the end for which we choose between them. For to study these subjects together will easily enable you to understand the causes of the particular phenomena. And those who have not fully accepted this, in proportion as they have not done so, will be ill acquainted with these very subjects, nor have they secured the end for which they ought to be studied.

LETTER TO IDOMENEUS ❧

(tr. Robert Drew Hicks)

On this last, yet blessed, day of my life, I write to you. Pains and tortures of body I have to the full, but there is set over against these the joy of my heart at the memory of our happy conversations in the past. Do you, if you would be worthy of your devotion to me and philosophy, take care of the children of Metrodorus.

Last Will ~

(tr. Robert Drew Hicks from *Lives of the Eminent Philosophers* by Diogenes Laertius)

In this manner I give and bequeath all my property to Amynomachus, son of Philocrates of Bate and Timocrates, son of Demetrius of Potamus, to each severally according to the items of the deed of gift registered in the Metroon, on condition that they shall place the garden and all that pertains to it at the disposal of Hermarchus, son of Agemortus, of Mitylene, and the members of his society, and those whom Hermarchus may leave as his successors, to live and study in. And I entrust to my School in perpetuity the task of aiding Amynomachus and Timocrates and their heirs to preserve to the best of their power the common life in the garden in whatever way is best, and that these may help to maintain the garden in the same way as those to whom our successors in the School may bequeath it. And let Amynomachus and Timocrates permit Hermarchus and his associates to live in the house in Melite for the lifetime of Hermarchus.

And from the revenues made over by me to Amynomachus and Timocrates let them to the best of their power in consultation with Hermarchus make separate provision for the funeral offerings to my father, mother, and brothers, and for the customary celebration of my birthday on the tenth day of Gamelion in each year, and for the meeting of all my School held every month on the twentieth day to commemorate Metrodorus and myself according to the rules now in force. Let them also join in celebrating the day in Poseideon which commemorates my brothers, and likewise the day in Metageitnion which commemorates Polyaenus, as I have done previously.

And let Amynomachus and Timocrates take care of Epicurus, the son of Metrodorus, and of the son of Polyaenus, so long as they study and live with Hermarchus. Let them likewise provide for the maintenance of Metrodorus's daughters so long as she is well-ordered and obedient to Hermarchus; and, when she comes of age, give her in marriage to a husband selected by Hermarchus from among the members of the School; and out of the revenues accruing to me let Amynomachus and Timocrates in consultation with Hermarchus give to them as much as they think proper for their maintenance year by year.

Let them make Hermarchus trustee of the funds along with themselves, in order that everything may be done in concert with him, who has grown old with me in philosophy and is left at the head of the School. And when the girl comes of age, let Amynomachus and Timocrates pay her dowry, taking from the property as much as circumstances allow, subject to the approval of Hermarchus. Let them provide for Nicanor as I have done previously, so that none of those members of the school who have rendered service to me in private life and have shown me kindness in every way and have chosen to grow old with me in the School should, so far as my means go, lack the necessaries of life.

All my books to be given to Hermarchus.

And if anything should happen to Hermarchus before the children of Metrodorus grow up, Amynomachus and Timocrates shall give from the funds bequeathed by me, so far as possible, enough for their several needs, as long as they are well ordered. And let them provide for the rest according to my arrangements; that everything may be carried out, so far as it lies in their power. Of my slaves I manumit Mys, Nicias, Lycon, and I also give Phaedrium her liberty.

Vatican Sayings* 🐾
(tr. Peter Saint-Andre, 2010)

4. Pain is easily disdained; for a pain that causes intense suffering is brief, whereas a pain that lingers in the flesh is weak and feeble.

7. It is easy to commit an injustice undetected, but impossible to be sure that you have escaped detection.

9. Compulsion is a bad thing, but there is no compulsion to live under compulsion.

11. For most people, to be quiet is to be numb and to be active is to be frenzied.

14. We are born only once and cannot be born twice, and must forever live no more. You don't control tomorrow, yet you postpone joy. Life is ruined by putting things off, and each of us dies without truly living.

15. We treasure our character as our own, whether or not it is worthy in itself or admired by others; and so we must honour our fellow men, if they are good.

16. No one who sees what is bad chooses it willingly; instead he is lured into seeing it as good compared to what is even worse, and thus he is trapped.

* Not all these writings have survived. This text includes all the sayings that remain.

17. It is not the young man who is most happy, but the old man who has lived beautifully; for despite being at his very peak the young man stumbles around as if he were of many minds, whereas the old man has settled into old age as if in a harbour, secure in his gratitude for the good things he was once unsure of.

18. The passion of love disappears without the opportunity to see each other and talk and be together.

19. He who forgets the good things he had yesterday becomes an old man today.

21. Nature must be persuaded, not forced. And we will persuade nature by fulfilling the necessary desires, and the natural desires too if they cause no harm, but sharply rejecting the harmful desires.

23. Every friendship is an excellence in itself, even though it begins in mutual advantage.

24. Dreams have neither a divine nature nor a prophetic power; instead they come from the impact of images.

25. Poverty is great wealth if measured by the goals of nature, and wealth is abject poverty if not limited by the goals of nature.

26. Understand that short discourses and long discourses both achieve the same thing.

27. Whereas other pursuits yield their fruit only to those who have practised them to perfection, in the love and practice of wisdom knowledge is accompanied by delight; for here enjoying comes along with learning, not afterward.

28. Those who grasp after friendship and those who shrink from it are not worthy of approval; on the other hand, it is necessary to risk some pleasure for the pleasures of friendship.

29. Speaking freely in my study of what is natural, I prefer to prophesize about what is good for all people, even if no one will understand me, rather than to accept common opinions and thereby reap the showers of praise that fall so freely from the great mass of men.

32. Honouring a sage is itself a great good to the one who honours.

33. The body cries out to not be hungry, not be thirsty, not be cold. Anyone who has these things, and who is confident of continuing to have them, can rival the gods for happiness.

34. The use of friends is not that they are useful, but that we can trust in their usefulness.

35. Don't ruin the things you have by wanting what you don't have, but realize that they too are things you once did wish for.

37. Nature is weak in the face of what is bad, not what is good; for it is kept whole by pleasures and broken down by pains.

38. Anyone with many good reasons to leave this life is an altogether worthless person.

39. A friend is not one who is constantly seeking some benefit, nor one who never connects friendship with utility; for the former trades kindness for compensation, while the latter cuts off all hope for the future.

40. One who says that everything occurs by necessity cannot complain about someone who says that not everything occurs by necessity, because even that claim occurs by necessity.

41. One must laugh and seek wisdom and tend to one's home life and use one's other goods, and always recount the pronouncements of true philosophy.

42. At the very same time, the greatest good is created and the greatest evil is removed.

43. It is not right to love money unjustly, and shameful to love it justly; for it is unbecoming to be overly stingy, beyond what is right.

44. When the sage contends with necessity, he is skilled at giving rather than taking – such a treasury of self-reliance has he found.

45. The study of what is natural produces not braggarts nor windbags nor those who show off the culture that most people fight about, but those who are fearless and self-reliant and who value their own good qualities rather than the good things that have come to them from external circumstances.

46. We cast off common customs just as we would do to wicked men who have been causing great harm for a long time.

48. While you are on the road, try to make the later part better than the earlier part; and be equally happy when you reach the end.

52. Friendship dances around the world, announcing to each of us that we must awaken to happiness.

53. Envy no one. For good people do not deserve envy, and the more that wicked people succeed the more they ruin things for themselves.

54. Do not pretend to love and practise wisdom, but love and practise wisdom in reality; for we need not the appearance of health but true health.

55. Misfortune must be cured through gratitude for what has been lost and the knowledge that it is impossible to change what has happened.

56–57. The sage does not feel a greater pain when he is tortured than when his friend is tortured, and would die on his friend's behalf; for if he betrays his friend then the rest of his life would be troubled and disturbed on account of his treachery.

58. They must free themselves from the prison of public affairs and ordinary concerns.

59. The stomach is not insatiable, as most people say; instead the opinion that the stomach needs unlimited filling is false.

60. Everyone departs from life just as they were when newly born.

61. The sight of one's neighbours is most auspicious if it produces the like-mindedness of one's primary kin, or at least a serious interest in such like-mindedness.

62. If parents have cause to be angry with their children, of course it is foolish to resist, and thus not try to beg for forgiveness. But if they do not have cause and are angry without reason, it is ridiculous to make an appeal to one who is irrationally opposed to hearing such an appeal, and thus not try to convince him by other means in a spirit of good will.

63. There is an elegance in simplicity, and one who is thoughtless resembles one whose feelings run to excess.

64. The esteem of others is outside our control; we must attend instead to healing ourselves.

65. It is foolish to ask of the gods that which we can supply for ourselves.

66. We sympathize with our friends not through lamentation but through thoughtful attention.

67. A free person is unable to acquire great wealth, because that is not easily achieved without enslavement to the masses or to the powers that be. Instead, he already has everything he needs, and in abundance. But if by chance he should have great wealth, he could easily share it with his fellows to win their goodwill.

68. Nothing is enough to one for whom enough is very little.

69. The ingratitude of the soul makes a creature greedy for endless variation in its way of life.

70. Do nothing in your life which would cause you fear if discovered by your neighbour.

71. Ask this question of every desire: what will happen to me if the object of desire is achieved, and what if not?

73. Some bodily pains are worth enduring to ward off others like them.

74. In a scholarly dispute, he who loses gains more because he has learned something.

75. This saying is utterly ungrateful for the good things one has achieved: provide for the end of a long life.

76. I rejoice with you, for you are the kind of person I would praise if you were to grow old as you are, and who knows the difference between seeking wisdom for yourself and for the sake of Greece.

77. The greatest fruit of self-reliance is freedom.

78. The noble soul is devoted most of all to wisdom and to friendship – one a mortal good, the other immortal.

79. He who is at peace within himself also causes no trouble for others.

80. A young man's share in deliverance comes from watching over the prime of his life and warding off what will ruin everything through frenzied desires.

81. One will not banish emotional disturbance or arrive at significant joy through great wealth, fame, celebrity, or anything else which is a result of vague and indefinite causes.

GLOSSARY

ATOM: From a Greek word, atomos, meaning 'uncuttable'. Epicurus followed the Pre-Socratic atomists in maintaining that reality was fundamentally composed of atoms and 'void'. Unlike the early atomists, however, Epicurus allowed that atoms can be conceptually divided, so they have parts in that sense; since, however, they cannot be physically divided, they are indestructible, and thus atoms neither come into nor go out of existence. When he speaks of 'minimal parts' (cf. p. 79), he is speaking of conceptual, rather than physical parts – almost like points and lines in geometry, though he seems to have thought of the minimal parts as having some infinitesimally small size. This innovation was important for responding to criticisms of atomism (such as that, if they had no parts, they could have no shape).

EVIL (kaka): The Greek word need not be taken in a moral sense; in some cases, 'bad' may be less misleading as a translation.

HAPPINESS (eudaimonia): Happiness is a standard translation, but far from perfect. Eudaimonia is taken by ancient Greek moral philosophers to be the ultimate human good, at which all our actions aim, at least, insofar as we are rational. (For this reason, modern philosophers sometimes refer to Greek ethics generally as eudaimonistic.) What is important to bear in mind is that while the English word 'happiness' may carry a subjective connotation – one can be in a 'happy' mood or frame of mind – to live a eudaimōn life is to be doing objectively well. So, when Greek philosophers ask what eudaimonia consists in, they are asking what it is to live a successful, flourishing human life. This may bring out why Epicurus' identification of happiness with pleasure was more controversial than it may sound.

PLEASURE (hēdonē, from which derives our word 'hedonist'): Epicurus argued that highest good was the greatest pleasure – but he had a very peculiar conception of what the greatest pleasure was. To be free from pain and mental disturbance was, in itself, the greatest pleasure possible.

SELF-SUFFICIENCY/SELF-SUFFICIENT (autarkeia/ autarkēs): Along with most other Greek moral philosophers, Epicurus thought that a good, happy human life would be self-sufficient: Happiness involved being maximally immune from chance. This is perhaps part of the reason that he identified the greatest pleasure with tranquillity rather than sensual gratification. To pursue sensual gratification, one needs a fair amount of cooperation from the world; with some self-discipline, one can achieve tranquillity on one's own.

TELEOLOGY: Not a word used by Epicurus himself, but helpful in thinking about the differences among various philosophical systems. A teleological explanation is one that appeals to a telos – a purpose or goal, or related notion (as if I say that the heart beats because its purpose is to pump blood, or that lions have sharp teeth so that they can hunt their prey). One important feature of Epicurus' atomism, and any purely mechanistic system, is his rejection of natural teleology. According to Epicurus, unlike Aristotle, things do not happen in nature for the sake of some end, but because of the mechanical interactions among atoms.

TRANQUILLITY (ataraxia): By ataraxia, Epicurus roughly means 'freedom from (mental) disturbance or disorder'; 'calmness' and 'equanimity' are also used as translations. Epicurus identified ataraxia (along with aponia, freedom from pain) as the greatest pleasure, and thus with happiness. Other Hellenistic schools, such as the Stoics and the Sceptics, also took ataraxia as the ultimate human good, though they believed it was achieved in different ways.

VIRTUE (aretē): As with 'evil' for kaka, and 'happiness' for eudaimonia, the translation may be misleading in some contexts. The Greek word aretē need not have moralistic overtones, though in some cases it does. At the most general level, we might define aretē as the quality of a thing which makes it good of its kind (as does Plato in the Republic and Aristotle in the Nichomachean Ethics); for this reason, some translators prefer 'excellence'. When the ancient philosophers ask what the aretai of the human being are, they are asking what are the qualities that make someone excellent, which enable them to live a good human life. (In Homer, it is especially used of manly or martial qualities; by Epicurus' time, it was also used of cooperative virtues like friendliness and justice.) Though a hedonist, Epicurus hopes to salvage traditional virtues by arguing that possessing and displaying them conduces to pleasure in the long run.